Charles Augustus Smith

**Steam Making or Boiler Practice**

Charles Augustus Smith

**Steam Making or Boiler Practice**

ISBN/EAN: 9783743465121

Manufactured in Europe, USA, Canada, Australia, Japa

Cover: Foto ©berggeist007 / pixelio.de

Manufactured and distributed by brebook publishing software (www.brebook.com)

Charles Augustus Smith

**Steam Making or Boiler Practice**

# STEAM MAKING;

— OR —

# BOILER PRACTICE,

BY

### CHAS. A. SMITH, C. E.,

*Professor of Civil and Mechanical Engineering at Washington University, St. Louis, Mo.; Member of the American Society of Civil Engineers, the Engineers' Club of St. Louis, and Associate Member of the American Association of Railway Master Mechanics.*

CHICAGO:
THE AMERICAN ENGINEER, 182-184 DEARBORN STREET.
1885.

Entered according to act of Congress, in the year 1884, by
PROPRIETORS OF THE AMERICAN ENGINEER,
in the Office of the Librarian of Congress, at Washington, D. C.

29477

PRESS OF
JOHN W. WESTON,
CHICAGO, ILL.

# PUBLISHERS' PREFACE.

It is believed that some knowledge of the circumstances attending the publication of this work, "STEAM MAKING," as well as its companion volume, "STEAM USING," will be of interest to the reader.

The lamented author, Prof. Chas. A. Smith, had arranged with the *American Engineer* for the publication of the two works. While the first, "STEAM MAKING," was going through the columns of the *Engineer*, Professor Smith died, early in 1884, leaving also to the care of the *Engineer* the recently completed manuscript of "STEAM USING."

To all who are familiar with the circumstances under which the books were written—the author suffering from a mortal illness and struggling against death to thus round out his life work, only giving up to die on their completion—will appreciate and value the more highly the broad and active experience thus crystallized.

To Mr. John W. Weston, so long connected with this journal, and personally familiar with the author and his writings, has been delegated the pleasant duty of conducting these works through the various stages of bookmaking, with the result now presented. The task has not been without its difficulties, the most serious, perhaps, being the loss of the invaluable assistance of their author in the work of revision of matter and proof.

It has been the aim, as far as possible, to preserve the exact style of the author, and it is believed that the facts and features presented in both books, the heirlooms of an admirable man, acknowledged to be profound and exact in his particular lines of work, will be held to cover whatever defects of minor importance may be encountered.

THE AMERICAN ENGINEER.

CHICAGO, January 1, 1885.

## AUTHOR'S PREFACE.

In this work the author has aimed at the presentation of modern boilers and has intended to give only those sanctioned by general experience admitting nothing, for the sake of novelty or apparent advantages.

In such a work there can be little originality and the author is, of course, indebted to many sources for his information. The examples chosen are taken from American practice where possible. The four marine boilers and the boiler at Mulhouse are taken from "Engineering" as also the tables of experiments with rivetting and with the full sized boilers. To Mr. L. E. Fletcher the author is greatly indebted for permission to use his valuable paper on the Lancashire boiler in Chapter IV., and to H. K. Ivers, Assistant Engineer, United States Navy, for valuable assistance in preparing illustrations. Two good works on boilers have recently appeared in this country: The able work of Chief Engineer Shock, United States Navy, which treats mainly of the marine boiler, and that of Mr. Wm. Barr, of Indianapolis. The former is placed by its price beyond the reach of many, and the latter gives more attention to Western practice. In this book the attempt has been made to give a more comprehensive view of the ground than either of these authors, while necessarily many of the facts given are to be found in those works. On the practical construction of boilers nothing better has appeared than the work of Wilson, which is not, however, illustrated by examples in detail. The author would have wished this work to be illustrated much more fully, but for the desire to keep down the cost to a reasonable limit.

The table of experiments with boilers is compiled from all sorts of sources and is believed to be sufficiently extensive to furnish a precedent for almost any kind of boiler in any locality; however, nothing better than the deductions of Rankine and Clark have been made from them.

It was originally intended to include in this work the use of steam and to give an extended table of engine trials and a few examples of engines but the work has had to wait, and may follow this at some future time.

CHAS. A. SMITH.

# SKETCH OF THE LIFE AND CHARACTER OF THE AUTHOR.

Charles A. Smith was born in St. Louis, October 1, 1846. His parents were both Massachusetts people who had been still further west. From both father and mother he inherited the instincts of a sailor, and the blood of several generations of ship-masters coursed through his veins. Though he never became a sailor, he always showed a sailor's fondness for "fixing things," for using his hands, for actual construction.

While he was still an infant, his mother died of cholera in St. Louis, and he was placed in the care of his father's sister, in Newburyport, Mass. This kind aunt was his mother, and her house was his home till he had a home of his own. His mode of life was simple and plain, but young Smith made warm friends and his boyhood was happy.

I first met him in 1860, when I became principal of the Boys' High School, of Newburyport. He was then fourteen years old and a member of the second class. He was a pleasant little fellow with a frank, earnest look, and a forehead which suggested brains. When the school gave expression to its loyalty to the Union by the erection of a liberty pole and publicly celebrated a flag-raising, young Smith was selected by his schoolmates to mount the platform and haul home the stars and stripes.

The school had a very good theodolite, and when we came to Loomis' Surveying, a great enthusiasm for field work was developed, and young Smith was never so happy as when on a surveying party. He took the English course and graduated in 1862. The next spring he went into the office of J. B. Henck, civil engineer, in Boston. At that time he probably had no idea of going to an engineering school. In 1864 he was leveller on the Boston, Hartford & Erie Railway. In 1865 he became chief assistant in the City Engineer's office, Springfield, Mass. By this time he saw clearly that an engineer requires a training far beyond a high school education, and he resolved to enter the Massachusetts Institute of Technology, then first opened. He had been reading ahead somewhat, with occasional help from me, so that he entered what was organized as a sophomore class. He lived again in Newburyport and went eighty miles daily on his way to and from the Institute. President Rogers was his teacher in physics, Professor Runkle in mathematics and applied mechanics, and Professor Henck in civil engineering.

He graduated in the pioneer class in 1868. I never quite understood how he managed to meet the cost of his course at the Institute. To be sure he had carefully saved the earnings of three years, and he secured for his vacations most excellent employment under the celebrated hydraulic engineer, J. B. Francis, at Lowell, Mass. He there assisted in determining the flow of water in pipes, over wiers, the efficiency of turbines, etc. I left Massachusetts for St. Louis in 1865, so I did not follow closely his career as a student.

After a year as engineer on the Union Pacific Railway in Utah, he returned, on the completion of the road, to Boston and went into partnership with Professor J. B. Henck, as civil engineers. While there associated with Professor Henck, he took charge of a part of the Blue Ridge Railway of North Carolina, as division engineer.

At that time, in 1870, the steady development of the Polytechnic School of Washington University made it necessary to appoint an instructor of civil engineering. I took pleasure in recommending young Smith for the position, and he was appointed. For the first year he made his home in my family, and as a preparation for the work of the class room he read with me Rankine's Civil Engineering entire.

After a brief experience as instructor, Mr. Smith was appointed professor to the chair of civil and mechanical engineering, which was subsequently named in honor of William Palm. This chair Professor Smith held till June, 1883, when compelled by his last illness to resign.

Though devoted at all times to the work of his professorship, Professor Smith found time to mingle in matters of practical engineering. For five years he was consulting engineer of the Iron Mountain Railway, among other things designing the DeSoto shops, and building a new pier in the Black river. In a similar way he was associated with Messrs. Shickle, Harrison & Co., designing the arched ribs of the roof over the Chamber of Commerce, and the iron trestles of the Bessemer Iron Works. Professor Smith was engaged as consulting engineer for the construction of the water works of Hannibal, of St. Charles, in Missouri, and of Amesbury, Massachusetts. His last professional duties were in connection with the last named. The pumping works at Richmond, Va., were designed by him, his plans being entered in competition and receiving the first prize. In 1879 he spent his summer vacation as resident engineer of the Baltimore Bridge Company, building piers in the Mississippi river just below Minneapolis.

Without attempting to give a full list of the professional enterprizes of Professor Smith, I have said enough to show how tireless a worker he was, and how closely he studied the practical details of engineering. But it was in connection with the St. Louis Engineers' Club that his devotion and enthusiasm were most fully shown. He was an active member for twelve years, and the secretary for nine or ten years. The club has not always been as flourishing as it is now. It has had its seasons of depression when only the zeal and the courage of Secretary Smith seemed to hold it together. Nothing but the direst necessity compelled him to yield at last.

The fatal malady, which in the shape of a cancerous tumor, brought his life to an untimely close on the 2nd of February, 1884, was born, as he thought, of hard work, of exposure, and of physical neglect. He could scarcely stop to eat or sleep; it was work first and comfort last.

Nothing in Professor Smith's life was more heroic than the way he battled for two years against an impending fate. When too weak to stand before his class, he taught reclining upon a lounge. One of his last pupils speaks in a notice of his beloved professor of "the days of suffering spent in his study in the University, when we gathered round him as he lay on the lounge, unable to stand, and listened to his exposition of 'Economic Location,' taking as a basis the work of his friend, Arthur Wellington."

In January, 1883, he was forced to give up his class work altogether, and to keep his room. Still he was not idle. Lying on the bed, or reclining in an easy chair, he was hard at work upon his two books on "Steam Making" and "Steam Using," which are just now being issued by the *American Engineer*, in Chicago. The first was finished by the end of 1882, and arrangements were made for its publication, but the prospect for the second book was gloomy enough. Nevertheless, he worked at it with a terrible earnestness which no unfavorable symptom could diminish. Nay, though clinging to the faintest glimmer of hope of returning health, he toiled at his book with the resolute air of one who was fully conscious that his days were numbered, and that the book must speedily be finished. In spite of pain and the dark shadow of the inevitable, his mind seemed clear and his hand steady. In the spring of '83 he moved back to Newburyport, Mass., to be near his physician and his family friends. There in a quaint old house, in a quiet neighborhood of that quiet town, he finished his book, laying down his pen and the burden of life at the same time. The readers of "Steam Using" may be glad to know that the author's very life's blood went into that book; that it was the last, the most perfect fruit of a very active and noble life.

Professor Smith is a good example of a poor boy who made his own way; who fought his own battles; who earned and honored every position he took. He was always a student. Some of you will remember with what enthusiasm he studied quarternions and thermodynamics; with what zeal and success he read all that he could get on graphical statics, and how many important additions he suggested. The records of the St. Louis Club probably will show that Professor Smith has presented more papers than any other member, past or present.

As an engineer, Professor Smith was bold and trustworthy. His confidence was based upon sound theory and careful practice. He was skillful in preparing estimates and was always well informed both as regards the latest improvements in engineering, and the best methods of working the materials of construction.

These accomplishments added greatly to his value as an instructor of young engineers. His students were brought very close to engineering work. Though well read in theory, he loved to dwell on the details of practice. He never lost an opportunity to learn a new process, or to study a new machine. He used to tell how, while resident engineer on a road in

New England, he tried his hand on the engine of the construction train till he was able to "stoke" and to "drive."

Professor Smith left a wife and three children. During her husband's long and discouraging sickness, Mrs. Smith was better than a faithful nurse: she brought aid to his self-imposed labor, and hope and cheer to his fainting spirit. So well did she understand the nature of his work and his needs, and so helpful was the assistance she brought, that it is not too much to say that without her positive coöperation and encouragement the two books which he leaves behind would never have been finished.

I will not speak of personal losses. I prefer to feel that we all had much to be thankful for in Professor Smith, and the nearest had the most. Though dying in his thirty-eighth year, Professor Smith's memory may well be preserved. The world is certainly the better for his having lived in it.

C. M. WOODWARD,
*Dean Polytechnic School,*
*Washington University, St. Louis, Mo.*

St. Louis, December, 7, 1884.

# CONTENTS.

### CHAPTER I.

ON THE NATURE OF HEAT AND THE PROPERTIES OF STEAM:—

Heat—Thermodynamics—Ratios of Volume to Pressure: Regnault's Ratios—The Carnot Engine—Making Steam—Measurement of Heat Expended—Table: The Properties of Saturated Steam—Examples in Calculation of Heat Expended, Etc.—Table: Factors of Evaporation—Its Use—Table: Expansion and Density of Pure Water—Entrained Water and its Measurement............................................. 1— 14

### CHAPTER II.

ON COMBUSTION:—

Principles of Combustion—Evaporative Power of Fuels—Losses by Imperfect Combustion—Effects of Air on Combustion, Quantity Required, and Quality of Certain Coals—Loss of Heat by Radiation and Conduction—Height of Stack—Tables of Boiler Trials......................... 15— 46

### CHAPTER III.

EXTERNALLY FIRED STATIONARY BOILERS:—

Boilers, their Shapes and Classes—Specification for Boilers for Meier Iron Company—Setting—Boilers for Nova Scotia Iron Company—"French" Boiler Tried at Mulhouse—Water Tube Boilers—Boilers on Mississippi River Boats: the "Montana"—Specification of Boilers for La Clede Rolling Mills—Boilers for the St. Louis Lead and Oil Company—Specification for 60-inch Horizontal Tubular Steam Boiler—Upright Boilers. 47— 73

### CHAPTER IV.

INTERNALLY FIRED STATIONARY BOILERS:—

The Lancashire Boiler—Cornish Boiler at Dusseldorf—Specification for Galloway Boiler for Crystal Plate Glass Company........................ 74—100

## CHAPTER V.

INTERNALLY FIRED PORTABLE, LOCOMOTIVE AND MARINE BOILERS:—

PAGE.

Varieties of Boilers—Locomotive Boiler for Engine No. 150, Wabash, St. Louis & Pacific Railway—Boiler for "Consolidation Locomotive," Missouri Pacific Railroad—Marine Boilers—Boilers of H. M. S. "Rover"—Steel Boiler, Wallsend Slipway Co., Newcastle-Upon-Tyne—Boilers of S. S. "Assyrian Monarch"—Boilers for S. S. "Mexican"—Boilers of the Cunard Steamship "Servia"—Steam Fire Engine Boilers—The Herreshoff Boiler.................................................................101–119

## CHAPTER VI.

THE DESIGN, CONSTRUCTION AND STRENGTH OF BOILERS;—

The Special Features—United States Laws, Regulation and Inspection, Steam Pressure, Quality of Plates, Appurtenances—Table of Pressures Allowable—Table of Pressures, Etc., Allowable on Freight and Towing Vessels—Rules and Regulations Relating to Pressures, Boilers, and the Inspection of Boiler Plates—Laws Relating to Instruments—Discipline—Instruments, Machines and Equipments Approved for Use with Steamship Boilers—Extracts from English Board of Trade Rules—Rivetting—Experiments by David Greig and Max Eyth, and Conclusions—Conclusions of United States Board of Engineers on Bolts, in Plates—Eye Bars and Tie Stays—Strength of Flues.....................120–148

## CHAPTER VII.

DESIGN AND CONSTRUCTION CONTINUED—PROPORTIONS OF HEATING SURFACE, ETC.—ECONOMIC EVAPORATION—EXPLOSIONS:—

Experiments on the Collapsing of flues—Tests Made by the Manchester Steam Users' Association upon a Lancashire Boiler—Experiments by Messrs. John Elder & Company—Durability: Corrosion—Experiments on Steel by Wm. Boyd—Mr. W. Parker on Marine Steel Boilers—Mr. Robert Wilson's Conclusions on Water-Tube Boilers—Circulation of Water—On Heating Surface—Table of Desirable Efficiencies for Various Pressures and Times Between Coaling for Ordinary Marine Boilers, Ordinary Draft—Cost of Fuel, Etc.—Questions of Cost, Economy, Etc.—The Style of Boiler for Various Conditions, Etc.—Boiler Explosions—Explosion Experiments—Conclusions........................ ...........149–167

## CHAPTER VIII.

MISCELLANEOUS BOILERS—CHOICE OF BOILER FITTINGS AND APPURTENANCES:—

Fuels: Qualities, Quantities, Etc.—Quality of Water—Ogle's Boiler—Perkin's, Benson's, Belleville, Latta, and Herreshoff Boilers—Heine, Root and Firmenich Boilers—"Anthracite" Boiler—Kelly's, Shepherd's and the Harrison Boilers—Boiler Appurtenances—Feed-water Heaters—Feed Pumps—Injectors—Steam Blast—Blow-off Valves—Gauges—Safety Valves, Etc., Etc.........................................................168–195

# STEAM MAKING;
## —OR—
## BOILER PRACTICE.

#### CHAPTER I.
##### ON THE NATURE OF HEAT AND THE PROPERTIES OF STEAM.

By the term heat we understand that property of bodies by which they grow hot, and give the sensation with which we are all familiar.

Heat is produced in three ways:
   By chemical action,  A.
   By mechanical action, B.
   By electrical action,  C.

A.—When certain chemical elements or compounds are combined under certain circumstances, the result is a union accompanied by an increase of temperature and the development of heat; as for example, carbon or hydrogen combining with oxygen; sulphuric acid, or quick lime with water.

B.—By the mechanical work of friction or percussion: Examples of this are continually before us.

C.—By the passage of an electric current in a conductor,—as in wires of too great resistance; or the electric arc.

The property of heat is thought by some to consist of a kind of motion or vibration of the molecules of which bodies are supposed to consist;—for solid and liquid bodies in vibration, and for gaseous bodies in the real motion of the molecules. With the arguments, pro. or con., concerning this hypothesis we have little to do further than to state that, its truth appears very probable, and in such event the production of heat by chemical combination or the passage of an electric current is simply a kind of mechanical action; in the one case, the vibration resulting from the shock of molecules attracting each other; in the other, from the setting up of a wave movement, or kind of wave, in the path of the electric disturbance whatever that may be.

That heat was produced by mechanical means has been long known. While the identity of heat and mechanical force was suspected by Count Rumford nearly a hundred years ago, it was reserved for Joule to prove (by long continued experiment), that the same quantity of work always gave the same quantity of heat, and to Rankine and Clausius to show, theoretically that, the same quantity of heat always gives the same amount of work, which has since been proved beyond all doubt by experimental investigations.

By the labors of the two great men, Rankine and Clausius, the science of thermodynamics was created,—the application of mathematics it the laws of heat. Of this interesting and beautiful science we shall, however, only state the two fundamental principles:

*First Principle*—"Heat and mechanical energy are mutually convertible, "and heat requires for its production and produces by its disappearance "mechanical energy in the proportion of 772 foot-pounds for each British "unit of heat."

The British unit of heat, just mentioned, is: "The quantity of heat "which corresponds to an interval of one degree of Farenheit's scale in the "temperature of one pound of pure liquid water at and near its temperature "of greatest density (39.1°F)."

The second principle, as given by Clausius, is as follows:

*Second Principle.*—"Heat, of itself, never passes from a cold body to a hotter one."

Rankine states the second principle in a way that has been severely criticised by Maxwell, but which appears to mean that, a unit of heat in a cold body can do as much work as in a hot body, with the implied reservation that there must be yet a colder body into which it may pass.

Heat is converted into mechanical work through the agency of some body that is expanded by heat, such as air or water. The heat is transferred into these mediums, usually enclosed within limits of changeable volume, the expanding medium enlarging the volume against a resistance thereby does mechanical work.

It has been taken for granted that the word temperature was understood to have its ordinary meaning, and that neither the ordinary thermometric scales of temperature, nor the ordinary instruments used for measuring temperature required description; but when great accuracy was required, the use of the air thermometer drew attention to a very convenient scale. Dry air and some of the other gases increase in volume or pressure from the temperature of melting ice to that of boiling water under the atmospheric pressure as follows:

From the volume or pressure 1 to:

|  | Constant Volume. | Constant Pressure. |
|---|---|---|
| Air | 1.3665 | 1.3670 |
| Hydrogen | 1.3667 | 1.3661 |
| Nitrogen | 1.3668 | ...... |
| Carbonic Acid | 1.3688 | 1.3669 |
| Carbonic Oxide | 1.3667 | 1.3719 |
| Nitrous Oxide | 1.3676 | 1.3719 |
| Cyanogen | 1.3829 | 1.3877 |
| Sulphurous Acid | 1.3843 | 1.3903 |

NOTE.—The above ratios are from Regnault.

With the air thermometer the change in volume of a portion of dry air was used to measure the change in temperature, and the natural result was that the temperature at which the dry gas would have no volume, if the law should hold so far, was taken as the zero or starting point of such a scale. This zero is —461° F. or —273° C., and is called "absolute zero," and temperatures measured on this scale are called "absolute temperatures." We shall give later another and better reason for this scale and its name, for we know now that all the gases above given can be reduced to liquids and solids and therefore are not perfect gases.

A perfect or "reversible" engine was devised by Sadi Carnot; and although such an engine cannot be constructed, and if constructed, could not be worked; still it is extremely useful in assisting our conceptions and in giving us a limit beyond which we cannot hope to proceed with improvements.

The operation of the Carnot engine is as follows: From a hot body, at temperature $T_1$, a working body receives heat at the same temperature $T_1$, expanding and doing work from the heat in the hot body directly. After a time the hot body is withdrawn, leaving the working body at the same temperature $T_1$, and it then expands by virtue of the heat which it contains until its temperature has fallen to $T_2$. In expanding, more work has been effected, which, of course, goes to the credit of the engine as work done. At the temperature $T_2$, the working body is brought into contact with a body called the "cold body," at the same temperature $T_2$; work is then done on the working body from the outside in compressing it to such a point, heat meanwhile passing from the working body to the cold body at the same temperature. So that by continuing the process of compression after the removal of the cold body, the working body will have just reached its first state of volume, pressure and temperature; the work expended in the two compression processes is, of course, to the debit of the engine, but there is on the whole a balance of work done by the engine.

It can be shown in this case, whatever be the working substance used: First.—That this engine utilizes more heat than can be utilized by any other kind of engine working between the same temperatures $T_1$ and $T_2$. Second.—That the work done, or heat utilized, is to the heat expended from the hot body, as the difference between the temperatures between which the engine works, $T_1 - T_2$ is to the absolute temperature of the hot body $T_1$. Hence the fraction

$$\frac{T_1 - T_2}{T_1}$$

where $T$ is an absolute temperature, is known as the efficiency of the engine, and is the maximum efficiency which can be reached by theory.

The proof of the above statements is given in any work on thermodynamics, so that we shall not enter upon it here, believing it out of place in a work of a practical character.

From the properties of the Carnot engine, a scale of temperature, based upon the work done by a body when $T_1 - T_2 = 1°$, is established; and it has been shown that the scale thus established coincides in origin and amount with that of the perfect gas thermometer, which places it upon a more substantial basis.

When heat is put into any body it may either increase the agitation of its molecules, thereby heating it or raising its temperature; or it may expand it against an external resistance doing external work; or it may change its condition, overcoming molecular attractions, doing what is called internal work; or it may do two or three of these three things at the same time.

When a fire is lighted under a boiler containing cold water, the heat generated by the chemical action of combustion passes from the fire and the gaseous products of combustion to the iron of the boiler, through the iron of the boiler to the surface in contact with the water and thence into the water. The volume of the water slightly increases with the temperature, raising the level partly by its own increase in volume and partly by the increase in volume of the air contained in the water. The heat increases the molecular agitation of the water, till, usually at the temperature of 212° F., the boiler begins to make steam. If, as in many of the boiler trials, the man-head or safety valve is open; or, as in a common tea kettle, there is no other pressure than that of the air upon the water, at this temperature the water remains; and all the heat going into it is expended in overcoming the molecular attraction of one atom of water for another, and in forcing the molecules apart. In thus overcoming the molecular attraction it is doing internal work, and at the same time in lifting the atmosphere by the steam formed, it is performing external work.

When the quantities of heat which a pound of water requires to raise it from the temperature of melting ice into steam at any given pressure are measured, that which it takes to raise the temperature is not exactly the difference in the temperatures which would be required if the specific heat of water were constant, but a unit of heat raises the temperature of a pound of water a little less than one degree at the higher temperatures. When a boiler is making steam at a given pressure other than that of the atmosphere, there is a temperature at which steam forms from the water and above which the water cannot be raised. This is known as the temperature of evaporation for the pressure. It is to be noted that the pressure of the atmosphere may be partly removed and low pressure steam formed at less than atmospheric pressure.

The quantity of heat required to evaporate a unit of weight of water at different pressures, and to raise the temperature up to that of evaporation, was carefully determined by Regnault in an extensive series of experiments made at the expense of the French Government. The volume of one pound weight of steam, and, of course, its reciprocal, the density or weight of a cubic foot of steam, was determined by experiments made by Fairbairn and Tate.

From the heat of evaporation, the volume of steam, the pressure under which it was evaporated, and the volume of the water from which it was formed are computed:

*First.*—The external work in foot-pounds, or the product of the pressure in pounds per square foot by the difference in cubic feet of the volume of one pound of steam and one pound of water.

*Second.*—The external work in heat units obtained by dividing the external work in foot-pounds by 772.

*Third.*—The internal work of evaporation obtained by deducting from the heat of evaporation the external work found above.

*Fourth.*—The sum of the internal work of evaporation and the heat expended in raising the temperature,—sometimes called the total internal heat.

*Fifth.*—The sum of the heat expended in raising the temperature of the water, and the heat of evaporation; or, the sum of the total internal heat and the external work in heat units; or, the sum of the heat expended in raising the temperature, in internal work of evaporation and in external work, is called the total heat. These quantities may all be stated in foot-pounds, and some writers prefer to use them in this way. But, although the measurement of mechanical work is usually made in foot-pounds, all measurements of heat and steam which require measurements of temperature are best made with a thermometer, and by heat units; we shall, therefore, retain the heat units. There is also this advantage, that in computation there will be smaller numbers and less figures involved.

The measurement of the heat expended in raising the temperature of water, in the total internal heat and the total heat, are all based on a starting point of one pound of water at the temperature of melting ice. As, however, such quantities are usually used by differences, many writers give these data from 0° F. Of course this does not require any real existence to the imaginary pound of water, as water assumed in this way. It gives a little less numerical work with feed water at low temperature, but is of no help when the specific heat has varied so as to alter the heat expended in raising the temperature of the water from the difference between the temperature and 32°. We adhere to the basis of melting ice.

Most of the theoretical writers use as a base for the tables the temperature of evaporation, although others use the pressure,—a much more practical starting point for engineers. But these writers have not given the internal and external heats, have used in some cases the 0° F. starting point referred to above, and have given extended decimals. In our own table we have only given the nearest heat unit, and have given a table, not for every pound of pressure, it is true, but one in which it is very easy to interpolate the nearest unit. We believe this table to be convenient for use and sufficiently extended and accurate.

The heat of evaporation is called latent heat of evaporation, but as the term latent has now no meaning we shall not retain it.

As the Regnault experiments on steam are always considered models in every respect, and as being of unapproachable accuracy, we shall only say that they were made in all circumstances and conditions in a thoroughly practical way, and that the values reached have been computed from purely theoretical grounds; so also with densities. The table is to be relied upon, and we shall not explain the experiments or comment further upon them, but will illustrate by a few examples the use of the table here given:

STEAM MAKING; OR, BOILER PRACTICE.

## TABLE I.—THE PROPERTIES OF SATURATED STEAM.

| Pressure in lbs. per sq. inch above the atmosphere. | Temperature of steam in degrees Fahrenheit. | Heat above 32° F. in water at boiling point. | External work in heat units. | T'l heat above 32° Fahr. in steam. | Internal work of evaporat'n in heat units. | Latent heat of evaporat'n in heat units. | Total internal work above 32° in heat units. | Weight of 1 cu. ft. of steam in pounds. | Volume of 1 lb. in cubic feet. |
|---|---|---|---|---|---|---|---|---|---|
| −14 | 90 |  |  | 1109 |  |  |  |  |  |
| −13 | 121 | 99 | 62 | 1118 | 967 | 1029 |  | 0.006 | 172.0 |
| −12 | 138 | 106 | 65 | 1124 | 943 | 1018 |  | 0.008 | 117.5 |
| −11 | 150 | 113 | 67 | 1127 | 942 | 1009 |  | 0.011 | 89.6 |
| −10 | 160 | 128 | 67 | 1130 | 935 | 1002 |  | .014 | 72.6 |
| −9 | 168 | 136 | 67 | 1133 | 925 | 993 |  | .016 | 61.2 |
| −8 | 175 | 143 | 68 | 1134 | 923 | 991 |  | .019 | 52.9 |
| −7 | 181 | 150 | 68 | 1137 | 918 | 987 |  | .021 | 46.7 |
| −6 | 187 | 156 | 69 | 1138 | 913 | 982 |  | .024 | 41.8 |
| −5 | 192 | 161 | 69 | 1140 | 909 | 979 | As 1070 in practice. | .026 | 37.8 |
| −4 | 196 | 165 | 70 | 1141 | 906 | 976 |  | .029 | 34.6 |
| −3 | 201 | 170 | 70 | 1143 | 903 | 973 |  | .031 | 31.8 |
| −2 | 205 | 174 | 71 | 1144 | 899 | 970 |  | .034 | 29.5 |
| −1 | 209 | 178 | 71 | 1145 | 896 | 967 |  | .036 | 27.6 |
| 0 | 212 | 181 | 72 | 1146 | 893 | 965 | 1074 | .038 | 26.3 |
| 1 | 215 | 184 | 72 | 1147 | 890 | 962 | 1074 | .041 | 24.3 |
| 2 | 219 | 188 | 72 | 1148 | 888 | 960 | 1076 | .043 | 23.0 |
| 3 | 222 | 191 | 73 | 1149 | 887 | 958 | 1078 | .046 | 21.8 |
| 4 | 225 | 194 | 73 | 1150 | 885 | 956 | 1079 | .048 | 20.7 |
| 5 | 227 | 196 | 73 | 1151 | 882 | 953 | 1079 | .050 | 19.7 |
| 6 | 230 | 199 | 74 | 1152 | 879 | 951 | 1079 | .053 | 18.8 |
| 7 | 233 | 202 | 74 | 1152 | 877 | 950 | 1079 | .055 | 18.0 |
| 8 | 235 | 204 | 74 | 1153 | 876 | 948 | 1079 | .058 | 17.2 |
| 9 | 237 | 206 | 74 | 1154 | 873 | 947 | 1080 | .060 | 16.6 |
| 10 | 239 | 208 | 74 | 1154 | 872 | 945 | 1080 | .062 | 16.0 |
| 11 | 242 | 211 | 75 | 1155 | 869 | 944 | 1080 | .065 | 15.4 |
| 12 | 244 | 213 | 75 | 1156 | 867 | 942 | 1080 | .067 | 14.9 |
| 13 | 246 | 215 | 75 | 1156 | 866 | 941 | 1081 | .070 | 14.4 |
| 14 | 248 | 217 | 75 | 1157 | 864 | 939 | 1081 | .072 | 13.9 |
| 15 | 250 | 220 | 75 | 1158 | 863 | 938 | 1083 | .074 | 13.4 |
| 16 | 252 | 222 | 75 | 1158 | 862 | 937 | 1083 | .076 | 13.0 |
| 17 | 254 | 224 | 76 | 1159 | 859 | 935 | 1084 | .079 | 12.7 |
| 18 | 256 | 226 | 76 | 1159 | 858 | 934 | 1084 | .081 | 12.3 |
| 19 | 257 | 227 | 76 | 1160 | 857 | 933 | 1084 | .083 | 12.0 |
| 20 | 259 | 229 | 76 | 1160 | 856 | 932 | 1085 | .086 | 11.6 |
| 22 | 262 | 232 | 76 | 1161 | 853 | 929 | 1085 | .090 | 11.0 |
| 24 | 266 | 236 | 77 | 1162 | 850 | 927 | 1086 | .095 | 10.6 |
| 26 | 269 | 239 | 77 | 1163 | 848 | 925 | 1087 | .099 | 10.0 |
| 28 | 272 | 242 | 77 | 1164 | 846 | 923 | 1088 | .104 | 9.6 |
| 30 | 274 | 244 | 77 | 1165 | 844 | 921 | 1088 | .109 | 9.2 |
| 35 | 281 | 251 | 78 | 1167 | 838 | 916 | 1089 | .120 | 8.3 |
| 40 | 287 | 257 | 78 | 1169 | 834 | 912 | 1091 | .131 | 7.6 |
| 45 | 293 | 263 | 78 | 1171 | 830 | 908 | 1093 | .142 | 7.0 |
| 50 | 298 | 268 | 79 | 1172 | 825 | 904 | 1093 | .154 | 6.5 |
| 55 | 303 | 273 | 79 | 1174 | 822 | 901 | 1095 | .165 | 6.1 |
| 60 | 307 | 278 | 79 | 1175 | 818 | 897 | 1096 | .176 | 5.7 |
| 65 | 312 | 282 | 80 | 1176 | 814 | 894 | 1097 | .187 | 5.3 |
| 70 | 316 | 287 | 80 | 1178 | 811 | 891 | 1098 | .198 | 5.0 |
| 75 | 320 | 291 | 80 | 1179 | 808 | 888 | 1099 | .209 | 4.8 |
| 80 | 324 | 294 | 80 | 1180 | 806 | 886 | 1100 | .220 | 4.5 |
| 85 | 328 | 298 | 81 | 1181 | 802 | 883 | 1100 | .231 | 4.3 |
| 90 | 331 | 301 | 81 | 1182 | 800 | 881 | 1101 | .241 | 4.1 |
| 95 | 334 | 305 | 81 | 1183 | 798 | 878 | 1101 | .252 | 4.0 |
| 100 | 338 | 308 | 81 | 1184 | 795 | 876 | 1102 | .263 | 3.8 |

# NATURE OF HEAT AND PROPERTIES OF STEAM.

## TABLE I.—THE PROPERTIES OF SATURATED STEAM.

| Pressure in lbs. per sq. inch above the atmosphere. | Temperature of steam in degrees Fahrenheit. | Heat above 32° F. in water at boiling point. | External work in heat units. | T'l heat above 32° Fahr. in steam. | Internal work of evaporat'n in heat units. | Latent heat of evaporat'n in heat units. | Total internal work above 32° in heat units. | Weight of 1 cu. ft. of steam in pounds. | Volume of 1 lb. in cubic feet. |
|---|---|---|---|---|---|---|---|---|---|
| 105 | 341 | 311 | 82 | 1185 | 792 | 874 | 1103 | .274 | 3.6 |
| 110 | 344 | 315 | 82 | 1186 | 789 | 871 | 1104 | .284 | 3.5 |
| 115 | 347 | 318 | 82 | 1187 | 787 | 869 | 1105 | .295 | 3.4 |
| 120 | 350 | 321 | 82 | 1188 | 785 | 867 | 1106 | .306 | 3.3 |
| 125 | 353 | 324 | 82 | 1189 | 783 | 865 | 1107 | .316 | 3.2 |
| 130 | 355 | 327 | 82 | 1190 | 781 | 863 | 1108 | .327 | 3.1 |
| 135 | 358 | 329 | 82 | 1191 | 779 | 861 | 1108 | .338 | 3.0 |
| 140 | 361 | 331 | 83 | 1191 | 777 | 860 | 1109 | .348 | 2.9 |
| 145 | 363 | 334 | 83 | 1192 | 775 | 858 | 1109 | .359 | 2.8 |
| 150 | 366 | 337 | 83 | 1193 | 773 | 856 | 1110 | .369 | 2.7 |
| 155 | 368 | 340 | 83 | 1194 | 771 | 854 | 1111 | .380 | 2.6 |
| 160 | 371 | 341 | 83 | 1194 | 770 | 853 | 1111 | .390 | 2.6 |
| 165 | 373 | 344 | 83 | 1195 | 768 | 851 | 1112 | .400 | 2.5 |
| 170 | 375 | 347 | 84 | 1196 | 765 | 849 | 1112 | .412 | 2.4 |
| 175 | 377 | 348 | 84 | 1196 | 764 | 848 | 1113 | .422 | 2.4 |
| 180 | 380 | 351 | 84 | 1197 | 762 | 846 | 1113 | .433 | 2.3 |
| 185 | 382 | 353 | 84 | 1198 | 761 | 845 | 1114 | .443 | 2.3 |
| 195 | 386 | 357 | 84 | 1199 | 758 | 842 | 1115 | .463 | 2.2 |
| 205 | 390 | 361 | 85 | 1200 | 754 | 839 | 1115 | .484 | 2.1 |
| 215 | 394 | 365 | 85 | 1201 | 751 | 836 | 1116 | .505 | 2.0 |
| 225 | 397 | 368 | 85 | 1202 | 749 | 834 | 1117 | .525 | 1.9 |
| 235 | 401 | 373 | 85 | 1204 | 746 | 831 | 1119 | .546 | 1.8 |
| 245 | 404 | 376 | 85 | 1205 | 744 | 829 | 1120 | .567 | 1.8 |
| 255 | 408 | 380 | 85 | 1206 | 741 | 826 | 1121 | .587 | 1.7 |
| 265 | 411 | 383 | 85 | 1207 | 739 | 824 | 1122 | .608 | 1.6 |
| 275 | 414 | 386 | 85 | 1208 | 737 | 822 | 1123 | .627 | 1.6 |
| 285 | 417 | 389 | 86 | 1209 | 734 | 820 | 1123 | .649 | 1.5 |
| 335 | 430 | 392 | 86 | 1213 | 725 | 811 | 1127 | .750 | 1.3 |
| 385 | 445 | 417 | 86 | 1217 | 714 | 800 | 1131 | .850 | 1.2 |
| * * * | * * * | * * * | * * * | * * * | * * * | * * * | * * * | | |
| 435 | 457 | 428 | 87 | 1220 | 705 | 792 | 1133 | .950 | 1.05 |
| 485 | 467 | 440 | 87 | 1224 | 697 | 784 | 1137 | 1.049 | 0.95 |
| 585 | 487 | 460 | 87 | 1230 | 683 | 770 | 1143 | 1.245 | 0.80 |
| 685 | 504 | 477 | 88 | 1235 | 670 | 758 | 1147 | 1.439 | 0.69 |
| 785 | 519 | 493 | 88 | 1240 | 659 | 747 | 1152 | 1.632 | 0.61 |
| 885 | 534 | 507 | 88 | 1244 | 649 | 737 | 1156 | 1.823 | 0.55 |
| 985 | 516 | 520 | 88 | 1248 | 640 | 728 | 1160 | 2.014 | 0.50 |

Values below * * * are computed and not experimental.
NOTE.—For all values of Total Internal work below the atmosphere 1070 heat units may be taken. All decimal parts of heat units have been neglected and the last one may therefore be in error.

*Example I.*—How much more heat is needed to boil a pound of water at 200 pounds per square inch boiler pressure than at five pounds per square inch, the feed being at 60° F. in either case.

### AT FIVE POUNDS.
Units.
Heat required to raise 1 pound water from 32° to boiling at 5 pounds pressure....... 196
Deduct heat to raise from 32° to 60° not used.................................... 28

Heat to raise from 60° to boiling......................................... 168
Internal work of evaporation............................................. 882
External work of evaporation............................................. 73

Heat required to boil from feed at 60° at 5 pounds................................1,123

## AT TWO HUNDRED POUNDS.

|  | Units. |
|---|---|
| Heat required to raise 1 pound water from 32° to boiling at 200 pounds per sq. inch.. | 359 |
| Deduct heat to raise from 32° to 60° not used................................ | 28 |
|  | 331 |
| Internal work................................................... | 756 |
| External work................................................... | 84 |
|  | 1,171 |

Heat required to boil 1 pound of water from feed at 60° at 200 pounds:
1,171 − 1,123 = 48 units.

$$\frac{48}{1,123} = 4 \text{ per cent., nearly.}$$

The same result could be reached more directly.

|  | Units. |
|---|---|
| Total heat from 32° at 200 pounds.................................... | 1,199 |
| Total heat from 32° at 5 pounds...................................... | 1,151 |
| Difference............................................................ | 48 |

Deducting from the 1,151 the 28 units not used, from 32° to 60°, the feed being at 60°, we have 1,123 for the divisor to reduce to per cent. as before.

We advise the reader to use the former method, by preference, in his computations, as serving to keep in full view the different uses and the various amounts of heat required for them; although there is, of course, more numerical work required to do so.

The reason so much more difficulty is experienced in maintaining high pressure than low pressure steam is to be found, not in the boiling of equal weights of water, but in the fact that the high pressure steam leaves the boiler more easily. If, for example, it be employed in an engine, the engine can be made to do more work thereby. If, in running a boat, the boat going faster the engine uses more steam; if employed in heating a building, the radiators act more energetically with the higher pressure, transmit more heat, condense more steam, and the skillful attendant suits his fire to the work.

*Example II.*—How much saving of fuel can be made by raising the temperature of the feed-water from 100° F. to 200° F., the boiler pressure being 120 pounds per square inch.

|  | Units. |
|---|---|
| Total heat for 120 pounds.............................................. | 1,188 |
| Deduct in the one case the units not used in raising the water from 32° F. to 100° F... | 68 |
| Required from 100° F. to boil at 120 pounds............................ | 1,120 |
| In the other case deduct for not using from 32° to 200°................. | 169 |
| Required to boil at 120 pounds from water at 200° F.................... | 1,019 |

Difference between 1,120 and 1,019 is 101 units, or about 9 per cent.

In order to compare the performance of different boilers working with different pressures and fed with water at different temperatures, it is necessary to assume a standard pressure, temperature of evaporation, and temperature of feed-water. Various temperatures of feed-water have been used, 0° F., 32° F., 100° F., the latter about the usual temperature of feed-water for condensing engines, and 212° F., used more generally than any of the others as a standard; while for the pressure and temperature of evaporation the atmospheric pressure and 212° F. are usually taken.

*Example III.*—By experiment with a boiler at 160 pounds per square inch it was found that, one pound of coal evaporated 7.91 pounds of water. The temperature of the feed-water was noted at 120° F.: required the equivalent evaporation from and at 212° F.

|  | Units. |
|---|---|
| Total heat of evaporation from 32° F. at 160 pounds | 1,194 |
| Deduct from 32° to 120°, units not used | 88 |
| Heat to evaporate from 30° at 160 pounds | 1,106 |
| Internal heat of evaporation at 212° | 893 |
| External work of evaporation at 212° | 72 |
| Sum or heat of evaporation at 212° | 965 |

$$7.91 \times \frac{1,106}{965} = 9.06 \text{ as the evaporation required.}$$

In order to facilitate this computation the following table of factors of evaporation is given:

## TABLE II.—FACTORS OF EVAPORATION.

| Temperature of Feed-water in Degrees Fahr. | Pressure in Pounds per Square Inch above the Atmosphere. | | | | | | | | | | | | | | | | | | | | |
|---|---|---|---|---|---|---|---|---|---|---|---|---|---|---|---|---|---|---|---|---|---|
| | 0 | 5 | 10 | 15 | 20 | 25 | 30 | 35 | 40 | 45 | 50 | 60 | 70 | 80 | 90 | 100 | 120 | 140 | 160 | 180 | 200 |
| 32 | 1.187 | 1.192 | 1.195 | 1.199 | 1.201 | 1.204 | 1.206 | 1.209 | 1.211 | 1.212 | 1.214 | 1.217 | 1.219 | 1.222 | 1.224 | 1.227 | 1.231 | 1.234 | 1.237 | 1.239 | 1.241 |
| 35 | 1.184 | 1.189 | 1.192 | 1.196 | 1.198 | 1.201 | 1.203 | 1.206 | 1.208 | 1.209 | 1.211 | 1.214 | 1.216 | 1.219 | 1.221 | 1.224 | 1.228 | 1.231 | 1.234 | 1.236 | 1.238 |
| 40 | 1.179 | 1.184 | 1.187 | 1.191 | 1.193 | 1.196 | 1.198 | 1.201 | 1.203 | 1.204 | 1.206 | 1.209 | 1.211 | 1.214 | 1.216 | 1.219 | 1.223 | 1.226 | 1.229 | 1.231 | 1.233 |
| 45 | 1.173 | 1.178 | 1.181 | 1.185 | 1.187 | 1.190 | 1.192 | 1.195 | 1.197 | 1.198 | 1.200 | 1.203 | 1.205 | 1.208 | 1.210 | 1.213 | 1.217 | 1.220 | 1.223 | 1.225 | 1.227 |
| 50 | 1.168 | 1.173 | 1.177 | 1.180 | 1.182 | 1.185 | 1.187 | 1.190 | 1.192 | 1.193 | 1.195 | 1.198 | 1.200 | 1.203 | 1.205 | 1.208 | 1.212 | 1.215 | 1.218 | 1.220 | 1.222 |
| 55 | 1.163 | 1.168 | 1.171 | 1.175 | 1.177 | 1.180 | 1.182 | 1.185 | 1.187 | 1.188 | 1.190 | 1.193 | 1.195 | 1.198 | 1.200 | 1.203 | 1.207 | 1.210 | 2.213 | 1.215 | 1.217 |
| 60 | 1.158 | 1.163 | 1.166 | 1.170 | 1.172 | 1.175 | 1.177 | 1.180 | 1.182 | 1.183 | 1.185 | 1.188 | 1.190 | 1.193 | 1.195 | 1.198 | 1.202 | 1.205 | 1.208 | 1.210 | 1.212 |
| 65 | 1.153 | 1.158 | 1.161 | 1.165 | 1.167 | 1.170 | 1.172 | 1.175 | 1.177 | 1.178 | 1.180 | 1.183 | 1.185 | 1.188 | 1.190 | 1.193 | 1.197 | 1.200 | 1.203 | 1.205 | 1.207 |
| 70 | 1.148 | 1.153 | 1.156 | 1.160 | 1.162 | 1.165 | 1.167 | 1.170 | 1.172 | 1.173 | 1.175 | 1.178 | 1.180 | 1.183 | 1.185 | 1.188 | 1.192 | 1.195 | 1.198 | 1.200 | 1.202 |
| 75 | 1.143 | 1.148 | 1.151 | 1.155 | 1.157 | 1.160 | 1.162 | 1.165 | 1.167 | 1.168 | 1.170 | 1.173 | 1.175 | 1.178 | 1.180 | 1.183 | 1.187 | 1.190 | 1.193 | 1.195 | 1.197 |
| 80 | 1.137 | 1.143 | 1.146 | 1.149 | 1.151 | 1.154 | 1.156 | 1.159 | 1.161 | 1.162 | 1.164 | 1.167 | 1.169 | 1.172 | 1.174 | 1.177 | 1.181 | 1.184 | 1.187 | 1.189 | 1.191 |
| 85 | 1.132 | 1.137 | 1.140 | 1.144 | 1.146 | 1.149 | 1.151 | 1.154 | 1.156 | 1.157 | 1.159 | 1.162 | 1.164 | 1.167 | 1.169 | 1.172 | 1.176 | 1.179 | 1.182 | 1.184 | 1.186 |
| 90 | 1.127 | 1.132 | 1.135 | 1.139 | 1.141 | 1.144 | 1.146 | 1.149 | 1.151 | 1.152 | 1.154 | 1.157 | 1.159 | 1.162 | 1.164 | 1.167 | 1.171 | 1.174 | 1.177 | 1.179 | 1.181 |
| 95 | 1.122 | 1.127 | 1.130 | 1.134 | 1.136 | 1.139 | 1.141 | 1.144 | 1.146 | 1.147 | 1.149 | 1.152 | 1.154 | 1.157 | 1.159 | 1.162 | 1.166 | 1.169 | 1.172 | 1.174 | 1.176 |
| 100 | 1.117 | 1.122 | 1.125 | 1.129 | 1.131 | 1.134 | 1.136 | 1.139 | 1.141 | 1.142 | 1.144 | 1.147 | 1.149 | 1.152 | 1.154 | 1.157 | 1.161 | 1.164 | 1.167 | 1.169 | 1.171 |
| 105 | 1.111 | 1.117 | 1.120 | 1.123 | 1.125 | 1.128 | 1.130 | 1.133 | 1.135 | 1.136 | 1.133 | 1.141 | 1.143 | 1.146 | 1.148 | 1.151 | 1.155 | 1.158 | 1.161 | 1.163 | 1.165 |
| 110 | 1.106 | 1.111 | 1.114 | 1.118 | 1.120 | 1.123 | 1.125 | 1.128 | 1.130 | 1.131 | 1.133 | 1.136 | 1.138 | 1.141 | 1.143 | 1.146 | 1.150 | 1.153 | 1.156 | 1.158 | 1.160 |
| 115 | 1.101 | 1.106 | 1.109 | 1.113 | 1.115 | 1.118 | 1.120 | 1.123 | 1.125 | 1.126 | 1.120 | 1.131 | 1.133 | 1.136 | 1.138 | 1.141 | 1.145 | 1.148 | 1.151 | 1.153 | 1.155 |
| 120 | 1.096 | 1.101 | 1.104 | 1.108 | 1.110 | 1.113 | 1.115 | 1.118 | 1.120 | 1.121 | 1.123 | 1.126 | 1.128 | 1.131 | 1.133 | 1.136 | 1.140 | 1.143 | 1.146 | 1.148 | 1.150 |
| 125 | 1.091 | 1.096 | 1.099 | 1.103 | 1.105 | 1.108 | 1.110 | 1.113 | 1.115 | 1.116 | 1.118 | 1.121 | 1.123 | 1.126 | 1.128 | 1.131 | 1.135 | 1.138 | 1.141 | 1.143 | 1.145 |
| 130 | 1.085 | 1.091 | 1.094 | 1.097 | 1.099 | 1.102 | 1.104 | 1.107 | 1.109 | 1.110 | 1.112 | 1.115 | 1.117 | 1.120 | 1.122 | 1.125 | 1.129 | 1.132 | 1.135 | 1.137 | 1.139 |
| 135 | 1.080 | 1.085 | 1.088 | 1.092 | 1.094 | 1.097 | 1.099 | 1.102 | 1.104 | 1.105 | 1.107 | 1.110 | 1.112 | 1.115 | 1.117 | 1.120 | 1.124 | 1.127 | 1.130 | 1.132 | 1.134 |
| 140 | 1.075 | 1.080 | 1.083 | 1.087 | 1.089 | 1.092 | 1.094 | 1.097 | 1.099 | 1.100 | 1.102 | 1.105 | 1.107 | 1.110 | 1.112 | 1.115 | 1.119 | 1.122 | 1.125 | 1.127 | 1.129 |
| 145 | 1.070 | 1.075 | 1.078 | 1.082 | 1.084 | 1.087 | 1.089 | 1.092 | 1.094 | 1.095 | 1.097 | 1.100 | 1.102 | 1.105 | 1.107 | 1.110 | 1.114 | 1.117 | 1.120 | 1.122 | 1.124 |
| 150 | 1.065 | 1.070 | 1.073 | 1.077 | 1.079 | 1.082 | 1.084 | 1.087 | 1.089 | 1.090 | 1.092 | 1.095 | 1.097 | 1.100 | 1.102 | 1.105 | 1.109 | 1.112 | 1.115 | 1.117 | 1.119 |
| 155 | 1.059 | 1.065 | 1.068 | 1.071 | 1.073 | 1.076 | 1.078 | 1.081 | 1.083 | 1.084 | 1.086 | 1.089 | 1.091 | 1.094 | 1.096 | 1.099 | 1.103 | 1.106 | 1.109 | 1.111 | 1.113 |
| 160 | 1.054 | 1.059 | 1.062 | 1.066 | 1.068 | 1.071 | 1.073 | 1.076 | 1.078 | 1.079 | 1.081 | 1.084 | 1.086 | 1.089 | 1.091 | 1.094 | 1.098 | 1.101 | 1.104 | 1.106 | 1.108 |
| 165 | 1.049 | 1.054 | 1.057 | 1.061 | 1.063 | 1.066 | 1.068 | 1.071 | 1.073 | 1.074 | 1.076 | 1.079 | 1.081 | 1.084 | 1.086 | 1.089 | 1.093 | 1.096 | 1.099 | 1.101 | 1.103 |
| 170 | 1.044 | 1.049 | 1.052 | 1.056 | 1.058 | 1.061 | 1.063 | 1.066 | 1.068 | 1.069 | 1.071 | 1.074 | 1.076 | 1.079 | 1.081 | 1.084 | 1.088 | 1.091 | 1.094 | 1.096 | 1.098 |
| 175 | 1.039 | 1.044 | 1.047 | 1.051 | 1.053 | 1.056 | 1.058 | 1.061 | 1.063 | 1.064 | 1.066 | 1.069 | 1.071 | 1.074 | 1.076 | 1.079 | 1.083 | 1.086 | 1.089 | 1.091 | 1.093 |
| 180 | 1.033 | 1.039 | 1.042 | 1.045 | 1.047 | 1.050 | 1.052 | 1.055 | 1.057 | 1.058 | 1.060 | 1.063 | 1.065 | 1.068 | 1.070 | 1.073 | 1.077 | 1.080 | 1.083 | 1.085 | 1.087 |
| 185 | 1.028 | 1.033 | 1.036 | 1.040 | 1.042 | 1.045 | 1.047 | 1.050 | 1.052 | 1.053 | 1.055 | 1.058 | 1.060 | 1.063 | 1.065 | 1.068 | 1.072 | 1.075 | 1.078 | 1.080 | 1.082 |
| 190 | 1.023 | 1.028 | 1.031 | 1.035 | 1.037 | 1.040 | 1.042 | 1.045 | 1.047 | 1.048 | 1.050 | 1.053 | 1.055 | 1.058 | 1.060 | 1.063 | 1.067 | 1.070 | 1.073 | 1.075 | 1.077 |
| 195 | 1.018 | 1.023 | 1.025 | 1.030 | 1.032 | 1.035 | 1.037 | 1.040 | 1.042 | 1.043 | 1.045 | 1.048 | 1.050 | 1.053 | 1.055 | 1.058 | 1.062 | 1.065 | 1.068 | 1.070 | 1.072 |
| 200 | 1.013 | 1.018 | 1.021 | 1.025 | 1.027 | 1.030 | 1.032 | 1.035 | 1.037 | 1.038 | 1.040 | 1.043 | 1.045 | 1.048 | 1.050 | 1.053 | 1.057 | 1.060 | 1.063 | 1.065 | 1.067 |
| 205 | 1.008 | 1.013 | 1.015 | 1.020 | 1.022 | 1.025 | 1.027 | 1.030 | 1.032 | 1.033 | 1.035 | 1.038 | 1.040 | 1.043 | 1.045 | 1.048 | 1.052 | 1.055 | 1.058 | 1.060 | 1.062 |
| 210 | 1.008 | 1.008 | 1.011 | 1.015 | 1.017 | 1.020 | 1.022 | 1.025 | 1.027 | 1.028 | 1.030 | 1.033 | 1.035 | 1.038 | 1.040 | 1.043 | 1.047 | 1.050 | 1.053 | 1.055 | 1.057 |
| 212 | 1.002 | 1.002 | | | | | | | | | | | | | | | | | | | |

# NATURE OF HEAT AND PROPERTIES OF STEAM.

The use of this Table of Factors of Evaporation is readily seen by taking the last example. The boiler evaporating 7.91 pounds at 160 pounds per square inch from feed at 120° F., the evaporation factor from Table II. for 120° and 160 pounds is 1.146. 7.91 × 1.146 = 9.06, as before, for the equivalent evaporation from and at 212° F.

We introduce one other table here—the weight of 1 cubic foot of water at different temperatures. Very often in the trials of a boiler or engine the most convenient unit of measurement of water is the cubic foot. This will be the case when a weir measurement is made or when the water is measured by a water meter. The use of a water meter involves many precautions, the most important being the following: The meter should work under moderate head of supply and small head of delivery; it should be set in such a manner that it can be tested in place under the exact conditions of use; if a positive meter, it should be especially constructed to work freely, if it is to be used in warm water. This table is also used for estimating the weight of water in boilers, and for correcting boiler trials for differences of water level.

### TABLE III.—EXPANSION AND DENSITY OF PURE WATER.
FROM D. K. CLARK AND BY RANKINE. APPROXIMATE FORMULA.

| Temperature in degrees Fahrenheit. | COMPARATIVE. | | Density of Weight per Cubic Foot. |
|---|---|---|---|
| | Volume. | Density. | |
| 32   | 1.00000 | 1.00000 | 62.418 |
| 35   | 1.99993 | .00007  | 62.422 |
| 39.1 | 1.99989 | .00011  | 62.425 |
| 40   | 1.99989 | .00011  | 62.425 |
| 45   | 1.99993 | .00007  | 62.422 |
| 46   | 1.00000 | 1.00000 | 62.418 |
| 50   | 1.00015 | .99985  | 62.409 |
| 52.3 | 1.00029 | .99971  | 62.400 |
| 55   | 1.00038 | .99961  | 62.394 |
| 60   | 1.00074 | .99926  | 62.372 |
| 62   | 1.00101 | .99899  | 62.355 |
| 65   | 1.00119 | .99881  | 62.344 |
| 70   | 1.00160 | .99832  | 62.313 |
| 75   | 1.00239 | .99771  | 62.275 |
| 80   | 1.00299 | .99702  | 62.232 |
| 85   | 1.00379 | .99622  | 62.182 |
| 90   | 1.00459 | .99543  | 62.133 |
| 95   | 1.00554 | .99449  | 62.074 |
| 100  | 1.00639 | .99365  | 62.022 |
| 105  | 1.00739 | .99260  | 61.960 |
| 110  | 1.00889 | .99199  | 61.868 |
| 115  | 1.00989 | .99021  | 61.807 |
| 120  | 1.01139 | .98874  | 61.715 |
| 125  | 1.01239 | .98808  | 61.654 |
| 130  | 1.01390 | .98630  | 61.563 |
| 135  | 1.01539 | .98484  | 61.472 |
| 140  | 1.01690 | .98339  | 61.381 |
| 145  | 1.01839 | .98194  | 61.291 |
| 150  | 1.01989 | .98050  | 61.201 |
| 155  | 1.02164 | .97802  | 61.096 |
| 160  | 1.02340 | .97714  | 60.991 |
| 165  | 1.02580 | .97477  | 60.843 |
| 170  | 1.02690 | .97380  | 60.783 |

## TABLE III.—CONTINUED.

| Temperature in degrees Fahrenheit. | COMPARATIVE. | | Density of Weight per Cubic Foot. |
|---|---|---|---|
| | Volume. | Density. | |
| 175 | 1.02906 | .97193 | 60.655 |
| 180 | 1.03100 | .97006 | 60.548 |
| 185 | 1.03300 | .96828 | 60.430 |
| 190 | 1.03500 | .96632 | 60.314 |
| 195 | 1.03700 | .96440 | 60.198 |
| 200 | 1.03889 | .96256 | 60.081 |
| 205 | 1.0414 | .9602 | 59.93 |
| 210 | 1.0434 | .9584 | 59.82 |
| 212 by formula. | 1.0444 | .9575 | 59.76 |
| 212 by measurement. | 1.0466 | .9555 | 59.64 |
| 230 | 1.0529 | .9499 | 59.36 |
| 250 | 1.0628 | .9411 | 58.78 |
| 270 | 1.0727 | .9323 | 58.15 |
| 290 | 1.0838 | .9227 | 57.59 |
| 298 | 1.0899 | .9175 | 57.27 |
| 338 | 1.1118 | .8994 | 56.14 |
| 366 | 1.1301 | .8850 | 55.29 |
| 390 | 1.1444 | .8738 | 54.54 |

The use of the table of the properties of steam is more frequent in the study of engine performance and indicator diagrams than of boiler performance, but there is an important point in determining the evaporation of a boiler in which it becomes of use.

As bubbles of steam formed on the hot iron of a boiler rise through the water to the surface, breaking and scattering spray, a portion of water thus thrown up into the steam room is carried along with the steam, and unless more heat be supplied to evaporate this water it increases the volume caused by the steam condensed in the pipes in the upper portion of the boiler. This water carried with the steam is said to be "entrained" with it and is called "priming" by many writers. When the proportion of water becomes so large as to be evident in the action of the engine or the exhaust, it is usually called by engineers "foaming." The amount of such water is increased if the water is dirty and covered with scum, or if grease and alkali combine to form a soap. The amount of water which can be carried by steam in suspension is very great, but depends somewhat upon the velocity of the current of steam; if the passages are large, and the flow of steam of moderate velocity the water has time to drop out of the steam by the action of gravity. In some cases the amount of water carried in weight has been known to be three times that of the steam carrying it, although usually it does not exceed 10 to 15 per cent.

The higher the pressure of steam the greater its density and the quieter, other things being equal, is the process of ebullition and the smaller the quantity of entrained water. The amount of water thrown up in spray is largely dependent on the circulation, being much diminished by improvements in that direction. The area of surface water in contact with the steam seems to be an important matter according to some authorities, but

## NATURE OF HEAT AND PROPERTIES OF STEAM.

as this varies very greatly without any apparent effect, we are not inclined to attribute much importance to it. A violent rush of steam close to the top of a body of water is to be avoided, as even a current of air would throw spray in such a case.

The accurate determination of the water entrained with steam is a matter of great difficulty and at the same time of great importance in the determination of the performance of boilers and engines.

Four methods have been devised to measure the amount of water entrained, and two of them have been used in practice.

The first method, that of M. G. A. Hirn, is the most used. It depends upon the amount of heat given out by a known weight of a mixture of steam and water and is best performed as follows:

A barrel is set on a platform scale and a known weight of water run into it. It is convenient to put in 298 lbs. of water. Steam is taken from the top of the steam pipe by a rubber hose terminated by an iron pipe capped on the lower end and perforated with holes drilled obliquely to the radii, but in the plane thereof. This pipe is placed in the barrel of water and steam turned on; the scale is loaded 2 ℔s. more, and as the steam comes into the water the fluid increases in weight, and when the beam tips there is 300 lbs. of water. The temperature of the water is then carefully noted. The disposition of the jets keeps the water stirred up thoroughly, and the flow of steam into the water being horizontal only, the water remains steady. The weight is then increased 10 lbs., and when the the scale tips at 310 pounds the temperature is noted.

The number of heat units given to the water, in the barrel, by the steam and water from the boiler, is found by multiplying the 300 ℔s. of water by the rise in its temperature.

The portion which was dry steam gives up its internal heat of evaporation in condensing, and the external work done by the air upon the fluid in compressing it from steam to water, together make the latent heat of evaporation; and the whole fluid then falls in temperature from that due to the pressure in the boiler to the final temperature of the barrel.

Deducting from the heat gained by the water in the barrel, ten times the difference between the boiler temperature and the final temperature in the barrel, and dividing the remainder by ten times the latent heat at the boiler pressure, the quotient will be the fraction of the whole which is dry steam.

It is easily seen that with any other weight the process would be the same; but in place of the ten we should use the number of pounds run in between the noting of the temperatures.

The preliminary 2 ℔s. is to provide for any water which might have collected in the hose or connections while standing, and to render the operation uniform.

Sometimes a coil of pipe as a surface condenser is used, and the steam which is condensed therein is kept separate from the condensing water; but great care has to be used to get all the water condensed out of the coil. The accuracy of this method is dependent upon the delicacy of weighing and the reading of the thermometer; in unskillful hands the results are sometimes astonishing.

The second method is to put into the feed water a quantity of sulphate of soda, and to draw from the boiler, at intervals, from the lower gauge cock a small amount of water, keeping this water by itself; also to draw from the steam, condensing either by a coil of pipe in water, or a small pipe in air, taking care to draw only water without steam, at the same intervals, keeping the one separate from the other. A chemical analysis defines the proportion of sulphate of soda in each portion, and a division of the proportion of sulphate of soda in the portion from the steam by the proportion in that from the water gives the proportion of water entrained,—the basis of the method being the fact that steam does not carry the sulphate of soda, this being only carried by the hot water entrained. This method was used by Professor Stahlschmidt at the Dusseldörf Exhibition Boiler Trials.

A third method has been suggested: To enclose a portion of steam in a vessel placed inside the steam pipe, then closing it and removing it from the steam pipe, obtain the weight of the enclosed fluid, which, being in a known volume, the proportion of water can be found from the volume and density at the known pressure. There appear to be many practical difficulties in this method, and we are not aware that it has been used to any extent.

A fourth method is to have a small cylinder with piston enclosed in the steam, and to put a known volume of the cylinder in connection with the steam; then closing the communication, pull out the piston (which, of course, passes through proper stuffing boxes into the air) until the pressure in the cylinder begins to lower,—the water contained evaporating at the pressure, until, after it has been evaporated the pressure begins to fall with increase of volume. The increase of volume at constant pressure divided by the final volume is the proportion of water carried. This method promises well, but we have no knowledge of its use.

Steam formed in the presence of water is always saturated, that is, it is at the same temperature as the water, and cannot be raised above that temperature until the water is all evaporated; but after this has been done, or if the steam be heated in a separate vessel, the temperature rises nearly $2°$ F. for each unit of heat added to a pound in weight, while the steam increases in volume at first not very closely, but afterwards very nearly as a perfect gas, or by $\frac{1}{455}$ part of itself for each degree F. The amount of heat required to raise 1 lb. weight of dry steam $1°$ F., is stated as 0.47 of a unit, and 0.5 by different authorities, the first including Rankine, and the second Hirn. Steam thus raised in temperature is said to be superheated, but our knowledge of this condition is still very limited and confined to the results of a few experiments.

# CHAPTER II.

## ON COMBUSTION.

The process of combustion is well known to be due to the act of uniting carbon and hydrogen with oxygen: other substances, such as sulphur and phosphorous also develop heat when uniting with oxygen, but for our practical purposes, carbon and hydrogen only need to be considered. In fact, hydrogen is a very important element in fuel, although forming but a very small part by weight of ordinary coal, the fuel most in use as a combustible.

The first question which arises is, how much air must be supplied to our fuel in order to produce complete combustion,—the air being required for the oxygen therein contained. The quantity of air required varies with the composition of the fuel, but if we say that for each pound of fuel we must supply twelve pounds of air, we shall be sufficiently near the truth. The volume of air will, of course, depend upon its temperature. Now, the quantity of heat which can be developed by the combustion of one pound of pure carbon, is sufficient to boil fifteen pounds of water from and at a temperature of 212° F. if none of the heat were lost; but there are many reasons why we do not reach this result in practice, and they are as follows:

*First.*—Variations in the quality of the coal as to its chemical constitution, affecting thereby its calorific power.

*Second.*—Impurities found with and mixed in the coal, affecting the actual quantity of pure coal in any given amount.

*Third.*—Imperfect or incomplete combustion of the fuel.

*Fourth.*—Losses of heat from the furnace, the fire, and the metal of the boiler.

*Fifth.*—The heat carried off in the stack, more or less utilized in the creation of draft.

1. *Variations in the Quality of Fuel.*—From the results of chemical analyses, the evaporative power of various kinds of fuel, expressed in pounds of water per pound of fuel evaporated from and at 212° F., which we will call $E$, have average values which are given in the following table:

KIND OF FUEL.

| | $E$. |
|---|---|
| Pure carbon completely burned to $CO_2$ | 15 |
| Pure carbon incompletely burned to $CO$ | 4.5 |
| $CO$ completely burned to $CO_2$ | 3.9 |
| Charcoal from wood, dry | 14 |
| Charcoal from peat, dry | 12 |
| Coke good, dry | 14 |
| Coke average, dry | 13.2 |
| Coke poor, dry | 12.3 |
| Coal, anthracite | 15.3 |
| Coal, dry bituminous, best | 15.0 |
| Coal, bituminous | 14 |

Coal, caking, bituminous, best.................................................................16
Coal, Illinois, (from four mines near St. Louis)...................................12
Lignite......................................................................................................12.1
Peat, dry..................................................................................................10
Peat with one-fourth water.................................................................. 7.5
Wood, dry............................................................................................... 7.25
Wood with one-fifth water................................................................... 5.8
Wood, best dry pitch pine....................................................................10
Mineral oils, about................................................................................22.6

As to what can be practically obtained under favorable conditions, the table of Boiler Trials, at the close of this chapter, can answer for itself; and in most cases, the results given are the best that can be obtained with clean boilers and skillful firing. For ordinary service results from 75 to 80 per cent. of those given in the table may safely be counted upon.

2. Impurities in the coal being earthy matter, forms ashes in fires of low temperature, and slag or cinders in fires of high temperature; water is also present which has to be evaporated, forming steam, and even decomposing into hydrogen and oxygen, thereby absorbing heat which passes off from the furnace; in the latter case a re-combination may take place, whereby the heat of decomposition is given up, but that used in changing water into steam is lost by being carried off up the stack.

3. *Imperfect Combustion.*—Some coal is usually lost with the ashes by falling through the grate bars, especially with such kinds of coal as split in the fire. In some cases this is prevented by wetting the small coal, thus holding it together till when on the fire it swells and cakes by the heat; it is, however, doubtful if this remedy is an economical one. The amount of this and the preceding loss may in practice be inferred from the column headed, Percentage of Refuse in the table of Boiler Trials, at the end of this chapter.

From this table it would appear that the refuse is: For the best soft coals from 3 to 10 per cent., and for the Illinois coals from 10 to 20 per cent. From coal near St. Louis we have usually found nearly 12½ per cent., or one-eighth. For the anthracites from 10 to 20 per cent.

Taking all things together, we find in practice that the best coals are the English and Pittsburgh soft coals; next in value the anthracites, which are only inferior by reason of their greater proportion of refuse, and the results are nearly the same for the best soft coals and anthracites. The Illinois coal near St. Louis is 80 per cent. in theory, but has rarely been found in practice to exceed 67 per cent. of the best coals.

Wood has about half the evaporative power of coal, and the usual comparison is to rate one cord, 128 cubic feet, equal to one ton of coal. The wood is supposed to be dry hard wood or pitch pine and weighs about two tons. This is the practice of the master mechanics in this country in rating fuel in locomotives.

Indian corn has sometimes been burned and found when dry to be about equal to the same weight of wood. Corn cobs have been found to be equal to one-third by weight of Illinois coal, or say one-fourth of good coal, or one-half of good wood by weight.

Incomplete combustion produces a very great loss, and this is best explained by a quotation from Rankine's Steam Engine, p. 270:

"The burning of carbon is always complete at first, that is to say, one "pound of carbon combines with two and two-thirds pounds of oxygen, and "makes three and two-thirds pounds of carbonic acid, and although the "carbon is solid immediately before the combustion, it passes during the "combustion into the gaseous state, and the carbonic acid is gaseous. This "terminates the process when the layer of carbon is not so thick and the "supply of air not so small, but that oxygen in sufficient quantity can get "direct access to all the solid carbon. The quantity of heat produced is "14,500 thermal units, as already stated."

"But in other cases part of the solid carbon is not supplied directly "with oxygen, but is first heated and then dissolved into the gaseous state "by the hot carbonic acid gas from the other parts of the furnace. The "three and two-thirds pounds of carbonic acid from one pound of carbon "are capable of dissolving an additional pound of carbon, making four and "two-thirds pounds of carbonic oxide gas, and the volume of this gas, "is double that of the carbonic acid gas which produces it."

"In this case the heat produced, instead of being that due "the complete combustion of one pound of carbon = heat units. 14,500 "falls to the amount, due to the imperfect combustion of two "pounds of carbon, or 2 × 4400, = heat units...................... 8,800

"Showing a loss of heat to the amount of ........................ 5,700 "heat units, which disappears in volatizing the second pound of carbon. "Should the process stop here as it does in furnaces ill supplied with air, "the waste of fuel is very great. But when the four and two-thirds "pounds of carbonic oxide gas containing two pounds of carbon, is mixed "with a sufficient supply of fresh air, it burns with a blue flame combining "with an additional two and two-thirds pounds of oxygen, making seven "and one-third pounds of carbonic acid gas, and giving additional heat of "double the amount due to the combustion of one and one-third pounds of "carbonic oxide. That is to say, 10,100 × 2 = heat units.......... 20,200 "To which add the heat produced by the imperfect combustion of "two pounds of carbon............................................ 8,800

"There is obtained the heat due to the complete combustion of two "pounds of carbon 2 × 14,500 = heat units....................... 29,000

With coal that has little flame, a thin fire, with exactly the right draft, has been found to give the best results, producing exactly the effects in the first part of the quotation.

It may be doubted if such a bad state of affairs is often found in a boiler furnace of the present day as indicated in the middle of the quotation, though a tendency to an insufficient supply of air may exist in internally fired boilers, such as locomotives, if there is a very thick fire and no air admitted above the grate; and, although not approaching remotely the case when no carbonic acid is produced, some of the carbonic oxide may pass off unburned. In such cases the admission of air above the fuel will be found beneficial.

In all soft coals there are found compounds of carbon and hydrogen known as hydro-carbons, which must also pass into the gaseous condition before being burned. "If these hydro-carbons such as pitch, tar, naptha "etc., are mixed on first issuing from the coal with a large quantity of air, "these inflammable gases are completely burned with a transparent blue "flame, producing carbonic acid and steam, but if raised to a red heat "before being mixed with air enough they disengage carbon in fine powder "and the higher the temperature the more carbon they disengage. If this "disengaged carbon is cooled below the temperature of ignition before "coming in contact with oxygen it constitutes while floating in gas smoke, "and when deposited on solid bodies is soot. But if this disengaged "carbon is maintained at the temperature of ignition, and supplied with "oxygen sufficient for its combustion, it burns while floating in the inflam- "mable gas with a red, yellow, or white flame. The flame from fuel is the "larger the more slowly its combustion is effected," and with the colors of flame given above as the combustion of smoke is less or more complete. An example of this is found in the use of common illuminating gas when burned with a "Bunsen" or a common burner. The chilling of the gaseous hydro-carbons, which are driven off from the solid pieces of coal by the heat developed, may take place in two ways: either by coming into contact with a cold body as the iron of the boiler, or by finding too much cold air in the furnace. To fully sustain the latter statement only a little consideration need be given to some of the fundamental principles of heat. It is well known that, if a certain amount of heat communicated to a body of certain weight and given material raises its temperature a definite number of degrees thereby, the same amount of heat communicated to twice the weight of the same material will only raise its temperature one-half the number of degrees that it was in the first case.

To apply this to combustion: One pound of carbon burned with twelve pounds of air gives thirteen pounds of gas at a temperature of 4580° F. above that of the external air; but it is found that this rarely, if ever, happens, and that to supply oxygen in plenty to the hot carbon surrounded by gas from 50 to 100 per cent. more air is used, and the result is from nineteen pounds of gas at a temperature of 3215° F. to twenty-five pounds of gas at a temperature of 2440° F. above the external air; but if forty-eight pounds of air per pound of coal were admitted, the resulting temperature of the forty-nine pounds of gas would be about 1250° F. above the external air. With anthracite coal and coke, such a lowering of temperature is not accompanied by serious loss, but with bituminous and semi-bituminous coals, such a reduction of the temperature of the fire is always productive of great waste.

To examine this more closely, suppose a coal with one-half free carbon and one-half hydro-carbon set on fire by the heat. If such a coal were burned with twelve pounds of air per pound of coal the temperature of the gas before the hydro-carbon ignited would be 2440° above the air, and the hydro-carbon would burn if supplied with oxygen enough and complete the combustion. Now if we burn this coal with twenty-four pounds of air per pound of coal, we have only about 1300° F., as temperature of the smoky product, and it is a question whether the gas would ignite; while with more air than this a great proportion of the gaseous fuel is lost and other evils are incurred.

We find then one marked point of difference between the anthracite and soft coals as fuel. While the former burns completely with a thin fire admitting an excess of air through it, and the free quantity of heat is developed, though the resulting temperature is not very high, the soft coal, on the contrary, absolutely requires for perfect combustion a high temperature and plenty of room before coming in contact with the iron of the boiler, and any deviation from these conditions produces smoke and great loss of heating power; and that while with hard coal too great a draft only wastes a small quantity of heat in the stack, with soft coal too great a draft may be as bad, or even worse, in its effects than too little.

With soft coal the required high temperature over the fire may be produced by intercepting the radiant heat of the fire by a fire brick arch or dome, which radiates back again to the fire, heating the products of combustion from both sides; this was first introduced by Mr. C. Wye Williams many years ago, and has been frequently revived in different forms since. In some devices air is introduced at the bridge, or at the edges of the arch or dome.

The great trouble with such arrangements has always been the lack of durability of the brick, used in the arch or dome. In fact, the more refractory the material the hotter the fire, and the destruction of the arch becomes only a question of, what is comparatively, a short time.

One of the satisfactory ways of obtaining a high temperature is by using so thick a bed of coal that the passage of too great a quantity of air is prevented by its friction upon the fuel: the thickness of fire being regulated by the size of the coal used, and kept so that it will not clinker too much. This effectually raises the temperature of the fire; it may also be done by the use of a damper, but not in so satisfactory a manner, although there is found to be in many cases a marked improvement by decrease in the draft. The general opinion in this country is decidedly in favor of thin fires, and the experiments of Professor Johnson at Washington favor this practice; but the experiments at Wigan, England, gave generally "the thicker the fire the better the result." Experiments with a pyrometer are needed in each case, but we may safely say that great improvement can be made in our practice in this respect, and that, the only secret in smoke prevention is to have a hot fire with room and time to let all the gas burn before coming to less than a red heat, and to fire in small quantities over a part of the grate at one time only.

Losses of heat by radiation and conduction from the furnace and ash pit of externally fired boilers are to be provided against by making the walls, if of brick, in two thicknesses with an air space between them; by keeping the ash pit doors partially closed, and by covering all radiating surfaces of metal with some good non-conducting material, such as thick felt faced on the inside with one-quarter inch of asbestos.

The amount of heat which may be lost by radiation from uncovered iron surfaces, exposed to air on one side and steam on the other, may be estimated as two and six-tenths heat units per square foot per hour per degree F. of difference of temperature between the steam and the air. If the air in the room be still, this amount may not be reached, but if exposed to violent winds it may be exceeded.

The heat passing up the chimney is not wholly lost, but is useful in producing a draft; and it can be shown that in a chimney where the draft is produced by the excess of weight of the outside air over that of the hot gas in the chimney, that the greatest quantity of gas by weight will pass up the chimney when the temperature of the gas in the chimney is about 625°F. hotter than the external air. With higher temperatures the velocity of flow will be greater and the quantity of gas by weight will be less owing to its greater volume. Looked at as a means of burning coal for making steam, the most coal that can be burned to advantage in a given time in a boiler furnace is when the temperature in the stack is near, but does not exceed, that of melting lead. A higher temperature than this means that the heat has not been properly taken out of the gas, and points to an increase in the boiler surface as a means of improving the performance of the boiler and increasing the yield of steam, as well as the economy of its production; a less temperature than the above is always desirable if the required quantity of steam can be maintained. In case twenty-four pounds of air per pound of fuel is used the temperature of stack giving maximum quantity of coal burned requires a little more than one-fourth of the heat generated to maintain the draft and the other three-quarters should pass into the water of the boiler. If we could get along with only twelve pounds of air per pound of fuel, only one-eighth of the heat generated would be required to maintain maximum draft. With forty-eight pounds of air per pound of fuel, one-half of the heat generated would be used in maintaining maximum draft. Here again the importance of hot fires is plainly indicated, and there is yet another reason for them: with a hot fire more of the heat generated passes into the water near the fire, leaving the products of combustion at a lower temperature to traverse the remainder of the surface and to leave the boiler at a lower temperature. More of the heat generated is therefore utilized than when the fire is not so hot.

A simple relation between the height of the stack in feet above the grate, its area in square feet, and the number of pounds of coal per minute burned, is the following equation, where:

$h$ = height in feet of the stack.
$A$ = area in square feet of stack.
$F$ = number of pounds of coal burned per minute.

$$h = \left(\frac{5F}{A}\right)^2 \qquad A = \frac{5F}{\sqrt{h}} \qquad F = \frac{A\sqrt{h}}{5}$$

It is understood, however, that $A$ is the "least flue area" in the passage of the hot gas.

## ON COMBUSTION.

The effects of changing the flue area, or as it is called the "calorimeter," and the proportions of heating surface and calorimeter to grate area are seen in the table of boiler trials following.

Gas has been employed as a fuel in boiler furnaces to a limited extent and for some years past, principally in Europe; but the knowledge of its adaptability, cleanliness, and heating qualities becoming wider, coupled with the discoveries of large reservoirs of natural gas in certain districts, has called closer attention to gas as a fuel and its use is largely extending. As data on this subject is limited, consequent upon the little knowledge extant of the results of the use of gaseous fuel, we do not consider it advisable to embody it in this work. In general cases there seems to be little chance of gain with properly constructed furnaces and with boilers of sufficient extent of heating surface to pay for the apparatus, simple as it is, of a producer.

The following tables, from page 22 to page 46, inclusive, give results of a large number of boiler trials, and need no further explanation.

## STEAM MAKING; OR, BOILER PRACTICE.

| Number for Reference | AUTHORITY | LOCATION | KIND OF BOILER | KIND OF FUEL | Grate area in square feet |
|---|---|---|---|---|---|
| 1 | B. F. Isherwood | Brooklyn W. W. | Return drop flue | Anthracite | 112.5 |
| 2 | " | " | " | " | |
| 3 | " | U. S. S. Michigan | Return water tube | " | 90 |
| 4 | " | " | " | " | |
| 5 | " | " | " | " | |
| 6 | " | " | " | " | |
| 7 | " | " | " | " | |
| 8 | " | " | " | " | |
| 9 | " | " | " | " | |
| 10 | " | " | " | Ormsby | |
| 11 | " | " | " | " | |
| 12 | " | " | " | " | |
| 13 | " | " | " | " | |
| 14 | " | " | " | Brookfield | |
| 15 | " | " | " | Anthracite | |
| 16 | " | U. S. S. Penguin | Return flue N. River type | " | 100 |
| 17 | " | U. S. S. Roanoke | Return water tube | " | 338 |
| 18 | " | U. S. S. Jacob Bell | 1 N. River | " | 47 |
| 19 | " | U. S. S. Bibb | 2 N. River | " | 84 |
| 20 | " | U. S. S. Mt. Vernon | 1 N. River | " | 79 |
| 21 | " | " | " | Semi-Bit | " |
| 22 | " | U. S. S. Valley City | " | Anthracite | 40 |
| 23 | " | " | " | " | |
| 24 | " | U. S. S. Crusader | " | " | 41 |
| 25 | " | " | " | " | |
| 26 | " | " | " | " | |
| 27 | " | U. S. S. Wyandotte | Return water tube | " | 59 |
| 28 | " | " | " | " | |
| 29 | " | " | " | " | |
| 30 | " | " | " | " | |
| 31 | " | " | " | " | |
| 32 | " | " | " | " | |
| 33 | " | " | " | Semi-Bit | " |
| 34 | " | U. S. S. Underwriter | 1 N. River | Anthracite | 76 |
| 35 | " | " | | Semi-Bit | " |
| 36 | " | U. S. S. Young America | " | Anthracite | 47 |
| 37 | " | Navy Yard, N. Y. | Return fire tube | " | 36 |
| 38 | " | " | " | " | |
| 39 | " | " | " | " | |
| 40 | " | " | " | " | |
| 41 | " | " | " | " | |
| 42 | " | " | " | " | |
| 43 | " | " | " | " | |
| 44 | " | " | " | " | |
| 45 | " | " | " | " | |
| 46 | " | " | " | " | |
| 47 | " | " | " | " | |
| 48 | " | " | " | " | |
| 49 | " | " | " | " | |
| 50 | " | " | " | Lackawanna | 36 |
| 51 | " | " | " | " | |
| 52 | " | " | " | Scranton | " |
| 53 | " | " | " | Boston | " |
| 54 | " | " | " | Hazelton | " |
| 55 | " | " | " | Pittston | " |
| 56 | " | " | " | Council Ridge | " |

ON COMBUSTION. 23

| Steam heating surface in sq. ft. | Least flue area in sq. ft. | Area of chimney. Sq. ft. | Height of chimney. ft. in. | Water heating surface ÷ grate area. | Steam heating surface ÷ grate area. | Grate area ÷ least flue area. | Grate area ÷ chimney area. | Lbs. coal per sq. ft. grate per hour. | Per cent. refuse. | EVAPORA- T'N FROM & AT 212°. Lbs. water ⑨ ℔. coal. | Lbs. water ⑨ ℔. com. | Duration in hours. | REMARKS. |
|---|---|---|---|---|---|---|---|---|---|---|---|---|---|
| 18.9 | 12.6 | 100 | 40 |  |  | 5.95 | 8.95 | 13.9 | 11 | 9.4 | 10.6 | 42 | |
| " | " | " | " |  |  | " | " | 13.5 | 11 | 9.3 | 10.5 | 43 | |
| 85 | 28.3 | 14.10 | 45 | 29.0 | 0.9 | 3.21 | 6.34 | 13.5 |  | 8.3 | 8.9 | 72 | |
| " | " | " | " | " | " | " | " | 11.4 |  | 8.2 | 10.1 | " | |
| " | " | " | " | " | " | " | " | 9.5 |  | 9.1 | 9.8 | " | |
| " | " | " | " | " | " | " | " | 6.3 |  | 9.6 | 10.2 | " | |
| " | " | " | " | " | " | " | " | 5.2 |  | 10.1 | 10.8 | " | |
| " | " | " | " | " | " | " | " | 3.8 |  | 9.4 | 9.9 | " | |
| " | " | " | " | " | " | " | " | 4.1 |  | 8.8 | 9.4 | " | |
| " | " | " | " | " | " | " | " | 4.4 | 6.0 | 9.4 | 10.0 | 296 | |
| " | " | " | " | " | " | " | " | 7.8 | 6.2 | 9.1 | 9.7 | 288 | |
| " | " | " | " | " | " | " | " | 12.2 | 5.5 | 9.0 | 9.5 | 252 | |
| " | " | " | " | " | " | " | " | 22.9 | 5.0 | 8.8 | 9.3 | 72 | |
| " | " | " | " | " | " | " | " | 18.5 | 6.8 | 8.3 | 8.9 | 72 | |
| " | " | " | " | " | " | " | " | 11.4 | 18.5 | 8.2 | 10.1 | 72 | |
| 262 | 13.5 | 10.8 | 51.6 | 27.0 | 2.6 | 7.32 | 5.95 | 11.9 | 23.3 | 8.7 | 11.4 | 72 | { 1 only in 4 used in experi- |
| .... | 53.5 | 50.3 | 61 | 35 | .... | 6.32 | 6.73 | 7.0 | 18.4 | 10.6 | 13.0 | 72 | } ment. |
| 183 | 6.1 | 7.0 | 50 | 31.4 | 4 | 7.7 | 6.7 | 11.4 | 14.2 | 10.4 | 12.1 | 72 | |
| 110 | 10.8 | 11.0 | 38.7 | 23.2 | 1.4 | 8.0 | 7.6 | 11.0 | 21.8 | 8.1 | 10.3 | 72 | |
| 276 | 12.4 | 12.6 | 55 | 30.1 | 5.2 | 5.3 | 6.3 | 11.7 | 10.0 | 9.5 | 10.5 | 49 | |
| " | " | " | " | " | " | " | " | 11.5 | 15.0 | 9.3 | 10.9 | 48 | |
| 180 | 4.9 | 8.7 | 34 | 23.8 | 4.5 | 8.2 | 4.6 | 11.0 | 17.0 | 9.4 | 11.3 | 48 | |
| " | " | " | " | " | " | " | " | 12.0 | 12.5 | 10.0 | 11.5 | 48 | |
| 8.6 | 5.1 | 7.1 | 43.5 | 25.6 | 2.1 | 8.1 | 5.8 | 10.7 | 8.7 | 10.3 | 11.8 | 96 | |
| " | " | " | " | " | " | 7.7 | " | 12.9 | 15.8 | 9.4 | 11.2 | 48 | |
| " | " | " | " | 17.51 | " | 12.8 | " | 12.9 | 19.8 | 9.0 | 11.2 | 48 | |
| .... | 10.0 | 9.6 | 50.9 | 37.2 | .... | 5.4 | 6.1 | 9.0 | 19.9 | 11.5 | 12.9 | 72 | |
| .... | " | " | " | " | " | " | " | 8.2 | 13.1 | 12.0 | 13.9 | " | |
| .... | " | " | " | " | " | " | " | 7.9 | 11.9 | 11.6 | 13.1 | " | |
| .... | " | " | " | " | " | " | " | 11.6 | 10.9 | 11.1 | 12.4 | " | |
| .... | " | " | " | " | " | " | " | 10.4 | 13.1 | 10.4 | 12.0 | " | |
| .... | " | " | " | " | " | " | " | 11.6 | 11.6 | 10.3 | 11.7 | " | |
| .... | " | " | " | " | " | " | " | 11.7 | 8.7 | 10.5 | 11.5 | " | |
| 327 | 12.5 | 12.6 | 52 | 23.9 | 4.0 | 9.5 | 6.0 | 11.2 | 11.4 | 10.0 | 11.3 | 48 | |
| " | " | " | " | " | " | " | " | 11.0 | 13.2 | 10.7 | 12.4 | 24 | |
| 37 | 6 0 | 7.9 | 47 | 20.8 | 0.3 | 7.8 | 6.0 | 11.3 | 17.2 | 8.7 | 10.5 | 72 | |
| " | 6.2 | " | 72 | 31.8 | " | 5.8 | " | 12.3 | 14.2 | 8.8 | 10.2 | 72 | 9 rows tubes. |
| " | " | " | " | " | " | " | " | 11.6 | 16.8 | 8.4 | 10.1 | " | 18 in a row, horizontal. |
| " | " | " | " | " | " | " | " | 13.9 | 17.3 | 9.3 | 11.3 | " | |
| " | 4.8 | " | " | 26.4 | " | 7.5 | " | 13.3 | 17.3 | 8.9 | 10.8 | 24 | 2 upper rows stop. |
| " | " | " | " | " | " | " | " | 12.4 | 20.0 | 8.3 | 10.4 | 48 | 2 lower rows stop. |
| " | " | " | " | " | " | " | " | 11.9 | 13.9 | 8.5 | 9.9 | " | 2 lower rows stop. |
| " | 4.1 | " | " | 23.4 | " | 8.8 | " | 9.8 | 17.3 | 9.8 | 11.8 | 50 | 3 upper rows stop. |
| " | " | " | " | " | " | " | " | 13.1 | 16.3 | 9.4 | 11.3 | 48 | 3 lower rows stop. |
| " | 3.4 | " | " | 21.1 | " | 10.5 | " | 10.5 | 18.1 | 9.9 | 12.1 | 48 | 4 upper rows stop. |
| " | " | " | " | " | " | " | " | 12.9 | 17.1 | 9.6 | 11.6 | 48 | 4 lower rows stop. |
| " | 4.6 | " | " | 25.8 | " | 7.8 | " | 12.4 | 19.1 | 8.5 | 10.5 | 48 | 2 vertical rows stop. |
| " | 3.8 | " | " | 22.8 | " | 9.4 | " | 13.2 | 14.0 | 9.4 | 10.9 | " | 3 vertical rows stop. |
| " | 3.1 | " | " | 19.8 | " | 11.7 | " | 12.8 | 16.9 | 9.3 | 11.2 | " | 4 vertical rows stop. |
| " | 6.2 | " | " | 31.8 | " | 5.8 | " | 12.6 | 17.5 | 9.2 | 11.2 | 73 | Anthracite. |
| " | " | " | " | " | " | " | " | 12.7 | 17.0 | 9.1 | 10.9 | " | |
| " | " | " | " | " | " | " | " | 12.2 | 17.2 | 8.4 | 10.1 | " | |
| " | " | " | " | " | " | " | " | 12.6 | 20.9 | 8.8 | 11.1 | " | |
| " | " | " | " | " | " | " | " | 13.0 | 15.0 | 8.9 | 10.4 | " | |
| " | " | " | " | " | " | " | " | 12.3 | 17.4 | 8.6 | 10.5 | " | |
| " | " | " | " | " | " | " | " | 12.6 | 10.8 | 9.7 | 10.8 | " | |

24　　　　　STEAM MAKING; OR, BOILER PRACTICE.

| Number for Reference. | AUTHORITY. | LOCATION. | KIND OF BOILER. | KIND OF FUEL. | Grate area in square feet. |
|---|---|---|---|---|---|
| 57 | B. F. Isherwood | Navy Yard, N. Y. | Return fire tube | Spring Mt. | 36 |
| 58 | " | " | " | Locust Mt. | " |
| 59 | " | " | " | Unknown | " |
| 60 | " | " | " | Broad Mt. | " |
| 61 | " | " | " | Black Heath | " |
| 62 | " | " | " | Broad Top | " |
| 63 | " | " | " | Cumberland | " |
| 64 | " | " | " | Eagleton | " |
| 65 | " | " | " | Glen Carbon | " |
| 65 | " | U. S. S. Monitor | 2 Return fire tube | Anthracite | 94 |
| 66 | " | " | " | " | " |
| 67 | " | U. S. Passaic | 2 Return water tube | " | 107 |
| 68 | " | U. S. Mackinaw | " | " | 200 |
| 69 | " | " | " | " | " |
| 70 | " | " | " | " | " |
| 71 | " | " | " | " | " |
| 72 | " | " | " | " | " |
| 73 | " | U. S. S. Eutaw | " | " | 80 |
| 74 | " | " | " | " | " |
| 75 | " | " | " | " | " |
| 76 | " | " | " | " | " |
| 77 | " | " | " | " | " |
| 78 | " | " | " | " | " |
| 79 | " | " | " | " | " |
| 80 | " | " | " | " | " |
| 81 | " | " | " | " | " |
| 82 | " | U. S. S. Gov. Buckingham | 1 N. River | " | 62 |
| 83 | " | " | " | " | " |
| 84 | " | " | " | " | |
| 85 | " | " | 1 N. River | " | 84 |
| 86 | " | U. S. S. Daylight | " | " | " |
| 87 | " | U. S. S. Shockokon | " | " | 65 |
| 88 | " | " | " | " | |
| 89 | " | U. S. S. Mahaska | Return water tube | " | 48 |
| 90 | " | U. S. S. Maratanza | " | " | 101 |
| 91 | " | U. S. S. San Jacinto | Fire tube | " | 108 |
| 92 | " | " | " | " | " |
| 93 | " | " | " | " | " |
| 94 | " | " | " | " | " |
| 95 | " | " | " | " | " |
| 96 | " | " | " | " | " |
| 97 | " | " | " | " | " |
| 98 | " | " | " | " | " |
| 66 | " | " | 1 Return water tube | " | " |
| 67 | " | " | " | " | " |
| 68 | " | " | " | " | " |
| 69 | " | " | " | " | " |
| 70 | " | " | " | " | " |
| 71 | " | U. S. S. Satellite | 1 N. River | " | 44 |
| 72 | " | U. S. S. Zouave | " | " | 32 |
| 73 | " | U. S. S. Com. Barney | " | " | 56 |
| 74 | " | U. S. S. Ella | " | " | 50 |
| 75 | " | U. S. S. Miami | Return fire tube | " | 49 |
| 76 | " | " | " | " | " |
| 77 | " | " | " | " | " |
| 78 | " | " | " | " | " |

## ON COMBUSTION.

| Steam heating surface in sq. ft. | Least flue area in sq. ft. | Area of chimney. Sq. ft. | Height of chimney. ft. in. | Water heating surface ÷ grate area. | Steam heating surface ÷ grate area. | Grate area ÷ least flue area. | Grate area ÷ chimney area | Lbs. coal per sq. ft. grate per hour. | Per cent refuse. | EVAPORATION FROM & AT 212°. | | Duration in hours. | REMARKS. |
|---|---|---|---|---|---|---|---|---|---|---|---|---|---|
| | | | | | | | | | | Lbs. water ⌀ lb. coal. | Lbs. water ⌀ lb. com. | | |
| 37 | 6.2 | 7.9 | 72 31.8 | 0.3 | 5.8 | 6.0 | | 11.1 | 13.8 | 9.2 | 10.0 | 73 | |
| " | " | " | " " | " | " | " | | 12.7 | 19.7 | 8.7 | 10.8 | " | |
| " | " | " | " " | " | " | " | | 12.3 | 14.1 | 8.8 | 10.2 | " | |
| " | " | " | " " | " | " | " | | 11.9 | 19.4 | 8.2 | 10.2 | " | |
| " | " | " | " " | " | " | " | | 13.9 | 17.3 | 9.3 | 11.3 | " | |
| " | " | " | " " | " | " | " | | 10.9 | 13.9 | 10.0 | 11.6 | " | Semi-bituminous. |
| " | " | " | " " | " | " | " | | 11.0 | 8.4 | 10.2 | 11.2 | " | Semi-bituminous. |
| " | " | " | " " | " | " | " | | 12.4 | 13.0 | 9.3 | 10.7 | " | Bituminous. |
| " | " | " | " " | " | " | " | | 14.0 | 17.7 | 9.2 | 11.2 | " | Bituminous. |
| 74 | 7.94 | 5.4 | 42.5 26.5 | 0.8 | 12.0 | 17.5 | | 6.0 | 24.0 | 9.6 | 12.8 | 72 | |
| " | " | " | " " | " | " | " | | 5.2 | 23.4 | 9.5 | 12.4 | 33 | |
| 35 | 13.3 | 8.7 | 27 32.9 | 0.3 | 8.0 | 12.3 | | 5.4 | 15.9 | 11.0 | 13.0 | 55 | |
| 171 | 26.7 | 23.8 | 58.5 25.2 | 0.8 | 7.5 | 8.4 | | 4.5 | 19.8 | 10.9 | 13.6 | 48 | |
| " | " | " | " " | " | " | " | | 6.3 | 18.3 | 10.1 | 12.4 | " | |
| " | " | " | " " | " | " | " | | 6.3 | 22.7 | 9.7 | 12.5 | " | |
| " | " | " | " " | " | " | " | | 5.3 | 17.6 | 10.6 | 12.8 | " | |
| " | " | " | " " | " | " | " | | 4.5 | 20.0 | 10.7 | 13.4 | " | |
| .... | 10.7 | 11.9 | " 23.5 | ... | " | " | | 8.4 | 19.1 | 10.4 | 12.8 | 72 | 4/5 part of one of two used, |
| .... | " | " | " " | .... | " | " | | 13.8 | 15.5 | 9.4 | 11.1 | " | No. 73 natural draft, others |
| .... | " | " | " " | .... | " | " | | 13.8 | 20.7 | 9.5 | 12.0 | " | steam jet. |
| .... | " | " | " " | .... | " | " | | 12.8 | 23.2 | 7.4 | 9.7 | " | |
| .... | " | " | " " | .... | " | " | | 4.8 | 24.0 | 9.3 | 12.2 | " | |
| .... | " | " | " " | .... | " | " | | 5.3 | 23.8 | 9.4 | 12.3 | " | |
| .... | " | " | " " | .... | " | " | | 8.5 | 21.6 | 10.2 | 13.1 | " | |
| .... | " | " | " " | .... | " | " | | 7.6 | 24.0 | 9.5 | 12.5 | " | |
| .... | " | " | " " | .... | " | " | | 9.8 | 23.1 | 9.7 | 12.6 | " | |
| 210 | 10.4 | 23.0 | 44 32.8 | 3.4 | 5.9 | 2.7 | | 13.1 | 16.7 | 9.0 | 10.8 | 48 | |
| " | " | " | " " | " | " | " | | 10.4 | 16.2 | 8.7 | 10.4 | " | |
| " | " | " | " " | " | " | " | | 9.4 | 16.0 | 9.0 | 10.7 | | |
| 110 | 7.0 | 8.3 | 43.6 30.4 | 1.3 | 12.0 | 10.2 | | 8.2 | 20.4 | 8.9 | 11.2 | 72 | |
| " | " | " | " " | " | " | " | | 10.4 | 20.4 | 9.4 | 11.9 | | |
| 195 | 7.4 | 13.6 | 55.9 21.1 | 3.0 | 8.8 | 4.7 | | 17.4 | 12.5 | 7.5 | 8.5 | 48 | |
| " | " | " | " " | " | " | " | | 10.1 | 12.5 | 7.5 | 8.5 | " | |
| 47 | 6.5 | 7.9 | 59.6 29.3 | 1.0 | 7.4 | 6.0 | | 6.7 | 17.4 | 11.5 | 13.9 | 72 | |
| 190 | 15.6 | 14.2 | 60.6 33.5 | 1.9 | 6.4 | 7.1 | | 9.6 | 26.2 | 9.7 | 13.1 | | |
| .... | 16.9 | 31.5 | 51.5 24.7 | .... | 6.5 | 6.8 | | 5.0 | 14.0 | 9.9 | 11.5 | " | With natural draft 1/6 only of |
| .... | " | " | " " | .... | " | " | | 12.6 | 14.3 | 8.9 | 10.4 | " | boiler for No. 91. |
| .... | " | " | " " | .... | " | " | | 10.9 | 19.0 | 9.5 | 11.7 | " | |
| .... | " | " | " " | .... | " | " | | 12.1 | 18.2 | 9.5 | 11.6 | " | |
| .... | " | " | " " | .... | " | " | | 11.3 | 19.0 | 9.9 | 12.4 | " | |
| ... | " | " | " " | .... | " | " | | 24.6 | 14.8 | 6.6 | 7.7 | " | With blower. |
| .... | " | " | " " | .... | " | " | | 23.3 | 16.0 | 7.0 | 8.3 | " | |
| .... | " | " | " " | .... | " | " | | 22.5 | 16.6 | 6.8 | 8.1 | " | |
| 448 | 16.9 | 11 | 51.5 30.5 | 4.1 | 6.4 | 6.9 | | 11.6 | 14.5 | 10.8 | 12.7 | " | Natural draft. |
| " | " | " | " " | " | " | " | | 11.2 | 18.0 | 10.6 | 12.9 | " | |
| " | " | " | " " | " | " | " | | 12.3 | 17.9 | 10.2 | 12.4 | " | |
| " | " | " | " " | " | " | " | | 10.5 | 20.4 | 10.7 | 12.4 | " | |
| " | " | " | " " | " | " | " | | 23.7 | 16.3 | 7.3 | 8.7 | " | Fan draft. |
| 212 | 5.0 | 7.1 | 49.2 27.3 | 4.8 | 8.7 | 8.1 | | 12.2 | 18.1 | 8.7 | 10.6 | " | |
| 32 | 2.5 | 3.1 | 33.7 21.3 | 1.0 | 12.8 | 10.3 | | 11.3 | 18.0 | 9.6 | 11.7 | " | |
| 231 | 6.3 | 8.7 | 50.0 27.3 | 4.1 | 8.9 | 7.9 | | 13.4 | 12.1 | 9.1 | 10.3 | " | |
| 152 | 5.3 | 5.1 | 51.5 24.2 | 3.1 | 9.8 | 9.2 | | 11.7 | 21.0 | 8.8 | 11.1 | " | |
| 48 | 6.9 | 7.1 | 50 24.2 | 1.0 | 7.1 | 7.0 | | 17.5 | 19.4 | 9.0 | 11.2 | 48 | |
| " | " | " | " " | " | " | " | | 17.7 | 19.9 | 8.9 | 11.1 | " | |
| " | " | " | " " | " | " | " | | 18.0 | 18.3 | 8.9 | 10.9 | " | |
| " | " | " | " " | " | " | " | | 19.0 | 17.7 | 8.8 | 10.7 | " | |

# STEAM MAKING; OR, BOILER PRACTICE.

| Number for Reference. | AUTHORITY. | LOCATION. | KIND OF BOILER. | KIND OF FUEL. | Grate area in square feet. | Water heating surface in sq. ft. |
|---|---|---|---|---|---|---|
| 79 | B. F. Isherwood | U. S. S. Miami | Return fire tube | Anthracite | 49 | 1198 |
| 80 | " | " | " | " | " | " |
| 81 | " | " | " | " | " | " |
| 82 | " | " | " | " | " | " |
| 83 | " | " | " | " | " | " |
| 84 | " | " | " | " | " | " |
| 85 | " | " | " | " | " | " |
| 86 | " | " | " | " | " | " |
| 87 | " | " | " | " | " | " |
| 88 | " | U. S. S. Philadelphia | " | " | 117 | 2941 |
| 89 | " | U. S. S. Dragon | N. River | " | 34 | 1007 |
| 90 | " | U. S. S. Gen. Putnam | " | " | 44 | 1118 |
| 91 | " | U. S. S. Whitehead | Return drop flue | " | 37 | 1038 |
| 92 | " | U. S. S. Morse | " | " | 56 | 1222 |
| 93 | " | N. Y. Navy Yard | Locomotive type | " | 5.3 | 115.4 |
| 94 | " | " | " | " | " | " |
| 95 | " | " | " | " | " | " |
| 96 | " | " | " | " | " | " |
| 97 | " | " | " | " | " | " |
| 98 | " | " | " | " | " | " |
| 99 | " | " | " | " | " | " |
| 100 | " | " | " | " | " | " |
| 101 | " | " | " | " | " | " |
| 102 | " | " | Return fire tube | " | 10.8 | 150 |
| 103 | " | " | " | " | " | " |
| 104 | " | " | " | " | " | " |
| 105 | " | " | " | " | " | " |
| 106 | " | " | " | " | " | " |
| 107 | " | " | " | Gas coke | " | " |
| 108 | " | " | " | Scotch Cannel | " | " |
| 109 | " | " | " | Anthracite | " | 125 |
| 110 | " | " | " | " | " | |
| 111 | " | " | " | " | " | 100 |
| 112 | " | " | " | " | " | |
| 113 | " | " | " | " | " | 75 |
| 114 | " | " | " | " | " | " |
| 115 | " | " | " | " | 8.6 | 149 |
| 116 | " | " | " | " | " | " |
| 117 | " | " | " | " | 6.5 | 148 |
| 118 | " | " | " | " | " | " |
| 119 | " | " | " | " | 4.3 | 147 |
| 120 | " | " | " | " | " | " |
| 121 | " | " | " | " | 10.8 | 150 |
| 122 | " | " | " | " | " | " |
| 123 | " | " | " | " | " | " |
| 124 | " | " | " | " | " | " |
| 125 | " | " | " | " | " | 45.5 |
| 126 | " | " | " | " | " | " |
| 127 | " | " | " | " | " | " |
| 128 | " | U. S. S. Kansas | Locomotive type | " | 54 | 1539 |
| 129 | " | " | " | " | " | " |
| 130 | " | " | " | " | " | " |
| 131 | " | " | " | " | " | " |
| 132 | " | U. S. S. Chippewa | " | " | 80 | 2919 |
| 133 | " | U. S. S. James Adger | " | " | 99 | 2060 |
| 134 | " | " | " | Cumberland | " | " |

ON COMBUSTION. 27

| Steam heating surface in sq. ft. | Least flue area in sq. ft. | Area of chimney. Sq. ft. | Height of chimney. ft. in. | Water heating surface ÷ grate area. | Steam heating surface ÷ grate area. | Grate area ÷ least flue area. | Grate area ÷ chimney area. | Lbs. coal per sq. ft. grate per hour. | Per cent. refuse. | EVAPORAT'N FROM & AT 212°. | | Duration in hours. | REMARKS. |
|---|---|---|---|---|---|---|---|---|---|---|---|---|---|
| | | | | | | | | | | Lbs. water ℔ lb. coal. | Lbs. water ℔ lb. com. | | |
| 48 | 6.9 | 7.1 | 50 | 24.2 | 1.0 | 7.1 | 7.0 | 17.7 | 17.4 | 8.9 | 10.7 | 48 | |
| " | " | " | " | " | " | " | " | 16.6 | 18.7 | 8.9 | 11.0 | " | |
| " | " | " | " | " | " | " | " | 16.1 | 18.8 | 9.1 | 11.3 | " | |
| " | " | " | " | " | " | " | " | 14.4 | 19.0 | 8.8 | 10.9 | " | |
| " | " | " | " | " | " | " | " | 17.0 | 19.0 | 8.7 | 10.8 | " | |
| " | " | " | " | " | " | " | " | 15.1 | 21.5 | 8.9 | 11.3 | " | |
| " | " | " | " | " | " | " | " | 13.2 | 21.4 | 8.5 | 10.8 | " | |
| " | " | " | " | " | " | " | " | 15.3 | 21.7 | 8.9 | 11.3 | " | |
| " | " | " | " | " | " | " | " | 11.8 | 16.4 | 9.1 | 10.9 | " | |
| 190 | 8.5 | 12.6 | 63 | 21.3 | 1.6 | 13.8 | 9.3 | 9.5 | 15.4 | 8.0 | 9.4 | 72 | |
| 65 | 4.2 | 4.9 | 35 | 29.6 | 1.9 | 8.1 | 6.9 | 10.1 | 11.8 | 10.3 | 11.6 | " | |
| 170 | 4.4 | 0.3 | 52 | 25.2 | 3.8 | 10.0 | 7.0 | 8.9 | 18.2 | 9.7 | 11.8 | " | |
| 70 | 3.1 | 5.9 | 37 | 27.7 | 1.9 | 12.0 | 6.3 | 4.6 | 25.0 | 12.1 | 13.5 | " | |
| 100 | 6.0 | 7.9 | 45.9 | 21.9 | 1.8 | 9.4 | 7.1 | 11.5 | 18.4 | 9.4 | 11.1 | " | |
| | 0.27 | 1.77 | | 16.6 | 21.9 | | 19.4 | 3.0 | 11.8 | 19.5 | 7.6 | 9.4 | 120 | |
| " | " | " | " | " | " | " | " | 10.9 | 19.6 | 7.7 | 9.5 | " | |
| " | " | " | " | " | " | " | " | 14.3 | 21.5 | 7.2 | 9.2 | " | |
| " | " | " | " | " | " | " | " | 15.9 | 18.8 | 7.4 | 9.1 | " | |
| " | " | " | " | " | " | " | " | 14.3 | 18.1 | 7.4 | 9.1 | " | |
| " | " | " | " | " | " | " | " | 14.4 | 19.2 | 7.5 | 9.2 | " | |
| " | " | " | " | " | " | " | " | 12.5 | 23.9 | 7.0 | 9.2 | " | |
| " | " | " | " | " | " | " | " | 20.5 | 21.6 | 7.4 | 9.5 | 72 | Uncovered. |
| " | " | " | " | " | " | " | " | 20.1 | 22.8 | 8.9 | 11.5 | " | Covered with felt. |
| 2.0 | 1.0 | 0.7 | 30.7 | 13.9 | 0.2 | 10.9 | 15.0 | 3.6 | 14.9 | 7.9 | 9.3 | 48 | |
| " | " | " | " | " | " | " | " | 11.0 | 14.7 | 7.6 | 9.0 | " | |
| " | " | " | " | " | " | " | " | 16.6 | 18.6 | 6.8 | 8.3 | " | |
| " | " | " | " | " | " | " | " | 22.1 | 17.7 | 6.7 | 8.1 | " | |
| " | " | " | " | " | " | " | " | 27.8 | 24.2 | 6.3 | 7.8 | " | |
| " | " | " | " | " | " | " | " | 16.6 | 8.6 | 7.6 | 8.3 | " | |
| " | " | " | " | " | " | " | " | 16.6 | 3.2 | 8.6 | 8.8 | " | |
| " | 0.04 | " | " | 11.6 | " | 14.5 | " | 16.6 | 22.1 | 6.3 | 8.1 | " | |
| " | " | " | " | " | " | " | " | 13.7 | 16.4 | 7.0 | 8.4 | " | |
| " | 0.49 | " | " | 9.2 | " | 21.8 | " | 16.6 | 20.0 | 7.0 | 8.7 | " | |
| " | " | " | " | " | " | " | " | 11.0 | 15.4 | 7.6 | 8.9 | " | |
| " | 0.25 | " | " | 6.9 | " | 43.6 | " | 16.4 | 24.9 | 6.7 | 8.0 | " | |
| " | " | " | " | " | " | " | " | 7.5 | 15.6 | 9.3 | 11.0 | " | |
| " | " | " | " | 17.2 | " | 8.7 | " | 15.1 | 19.6 | 7.2 | 9.1 | " | |
| " | " | " | " | " | " | " | " | 20.7 | 17.8 | 6.8 | 8.3 | " | |
| " | " | " | " | 22.8 | " | 6.5 | " | 14.9 | 22.9 | 7.1 | 9.2 | " | |
| " | " | " | " | " | " | " | " | 27.4 | 21.6 | 6.3 | 8.1 | " | |
| " | " | " | " | 34.0 | " | 4.4 | " | 15.4 | 21.4 | 7.6 | 9.7 | " | |
| " | " | " | " | " | " | " | " | 27.0 | 21.4 | 6.2 | 8.2 | " | |
| " | 0.74 | " | " | " | " | 14.3 | " | 10.5 | 10.3 | 6.8 | 8.6 | " | |
| " | 0.46 | " | " | " | " | 23.5 | " | 16.6 | 20.0 | 6.9 | 8.6 | " | Air holes above grate 0.4 sq.in. |
| " | 0.20 | " | " | " | " | 52.8 | " | 5.8 | 22.4 | 8.6 | 11.1 | " | per sq. ft. grate. |
| " | 1.0 | " | " | " | " | 10.9 | " | 16.5 | 16.4 | 7.0 | 8.4 | " | |
| " | 0.72 | " | " | 4.2 | " | 15.0 | " | 16.6 | 20.3 | 5.0 | 6.3 | " | Air as above. |
| " | 0.51 | " | " | " | " | 21.2 | " | 16.6 | 17.0 | 5.2 | 6.3 | " | No tubes out. |
| " | 0.25 | " | " | " | " | 43.4 | " | 11.8 | 27.1 | 5.7 | 7.8 | " | Chimney on back connection. |
| 40 | 9.9 | 8.7 | 50 | 28.5 | 0.7 | 5.5 | " | 17.0 | 10.9 | 8.5 | 10.2 | | |
| | 8.2 | " | " | " | " | 6.6 | " | 15.7 | 16.6 | 9.0 | 10.8 | | |
| | 9.6 | " | " | " | " | 8.5 | " | 16.1 | 18.2 | 8.8 | 10.8 | | |
| | 5.2 | " | " | " | " | 10.3 | " | 13.3 | 18.7 | 9.4 | 11.6 | | |
| 36 | 14.2 | 12.6 | 48 | 32.9 | 0.4 | 6.2 | " | 8.3 | 13.6 | 10.6 | 11.0 | | |
| 164 | 9.9 | 28.3 | 46 | 20.1 | 1.7 | 10.0 | " | 10.3 | 11.1 | 7.6 | 8.7 | | |
| " | " | " | " | " | " | " | " | 9.5 | 9.6 | 8.7 | 9.6 | | |

## 28　STEAM MAKING; OR, BOILER PRACTICE.

| Number for Reference. | AUTHORITY. | LOCATION. | KIND OF BOILER. | KIND OF FUEL. | Grate area in square feet. | Water heating surface in sq. ft. |
|---|---|---|---|---|---|---|
| 135 | D. K. Clark | Wigan, Eng. | 2 Lancashire | Bit. coal | 31.5 | 404 |
| 136 | " | " | " | " | " | 767 |
| 137 | " | " | " | " | " | " |
| 138 | " | " | " | " | " | " |
| 139 | " | " | " | " | " | " |
| 140 | " | " | " | " | " | " |
| 141 | " | " | " | " | " | " |
| 142 | " | " | " | " | " | " |
| 143 | " | " | " | " | " | 1314 |
| 144 | " | " | " | " | " | 1617 |
| 145 | " | " | " | " | " | " |
| 146 | " | " | " | " | 21 | 767 |
| 147 | " | " | " | " | " | " |
| 148 | " | " | " | " | " | " |
| 149 | " | " | " | " | " | " |
| 150 | " | " | " | " | " | " |
| 151 | " | " | 1 Galloway | " | " | 1617 |
| 152 | " | " | " | " | " | 431 |
| 153 | " | " | " | " | 31.5 | 719 |
| 154 | " | " | " | " | " | " |
| 155 | " | " | " | " | " | 1281 |
| 156 | " | " | " | " | " | 1569 |
| 157 | " | " | " | " | " | 719 |
| 158 | " | " | " | " | 21 | " |
| 159 | " | " | " | " | " | " |
| 160 | " | " | " | " | " | 1569 |
| 161 | " | " | Return fire tube | " | 10.3 | 508 |
| 162 | " | " | " | " | " | " |
| 163 | " | " | " | " | " | " |
| 164 | " | " | " | " | " | 301 |
| 165 | " | " | " | " | " | 503 |
| 166 | " | " | " | " | " | " |
| 167 | " | " | " | " | " | " |
| 168 | " | " | " | Newcastle | " | " |
| 169 | " | " | " | Wesh. | " | " |
| 170 | " | " | " | Lancashire. | " | " |
| 171 | " | " | " | Best Lancas're | " | " |
| 172 | " | " | " | Lowest " | " | " |
| 173 | " | Newcastle | Return fire tube | Hartley | 5.82 | 1069 |
| 174 | " | " | " | " | " | " |
| 175 | " | " | " | " | 19.2 | " |
| 176 | " | " | " | " | " | " |
| 177 | " | " | " | " | 22 | " |
| 178 | " | " | " | " | " | " |
| 179 | " | " | " | " | 18 | " |
| 180 | " | " | " | " | 15.5 | " |
| 181 | " | " | " | Newcastle | 42 | " |
| 182 | " | " | " | " | " | " |
| 183 | " | " | " | " | " | " |
| 184 | " | " | " | " | 33 | " |
| 185 | " | " | " | " | 22 | 749 |

## ON COMBUSTION.                                                                 29

| Steam heating surface in sq. ft. | Least flue area in sq. ft. | Area of chimney. Sq. ft. | Height of chimney. ft. in. | Water heating surface ÷ grate area. | Steam heating surface ÷ grate area. | Grate area ÷ least flue area. | Grate area ÷ chimney area. | Lbs. coal per sq. ft. grate per hour. | Per cent refuse. | EVAPORA-T'N FROM & AT 212° | | Duration in hours. | REMARKS. |
|---|---|---|---|---|---|---|---|---|---|---|---|---|---|
| | | | | | | | | | | Lbs. water ℔ lb. coal. | Lbs. water ℔ lb. com. | | |
| 164 | 9.9 | 21 | 90.9 15 | 1.7 | 10.0 | 15.0 | | 9.5 | 6.4 | 8.2 | 8.6 | 48 | Without side flues. |
| " | " | " | " 24.4 | " | " | " | | 18.6 | " | 10.3 | 11.0 | " | |
| " | " | " | " " | " | " | " | | 19.1 | " | 10.2 | 10.9 | " | |
| " | " | " | " " | " | " | " | | 20.1 | " | 9.6 | 10.3 | " | |
| " | " | " | " " | " | " | " | | 19.5 | " | 10.4 | 11.1 | " | |
| " | " | " | " " | " | " | " | | 17.6 | " | 10.6 | 11.3 | " | |
| " | " | " | " " | " | " | " | | 20.5 | " | 10.3 | 11.0 | " | |
| " | " | " | " " | " | " | " | | 21.0 | " | 10.3 | 11.0 | " | |
| " | " | " | " 41.4 | " | " | " | | " | " | 10.8 | 11.5 | " | } Without side flue, with economizer. " " " |
| " | " | " | " 51.3 | " | " | " | | 20.8 | " | 11.5 | 12.3 | " | " " " |
| " | " | " | " " | " | " | " | | 16.9 | " | 11.5 | 12.3 | " | |
| " | " | " | " 36.5 | " | " | " | | 22.7 | " | 10.9 | 11.6 | " | Without economizer. |
| " | " | " | " " | " | " | " | | 21.6 | " | 10.8 | 11.5 | " | |
| " | " | " | " " | " | " | " | | 23.0 | " | 10.6 | 11.3 | " | " " |
| " | " | " | " " | " | " | " | | 23.8 | " | 10.6 | 11.3 | " | |
| " | " | " | " 77 | " | " | " | | " | " | 11.6 | 12.4 | " | With economizer. |
| " | " | " | " 13.6 | " | " | " | | " | " | 8.5 | 9.0 | " | Without side flues. |
| " | " | " | " 22.8 | " | " | " | | 18.3 | " | 10.2 | 10.9 | " | |
| " | " | " | " " | " | " | " | | 20.0 | " | 9.7 | 10.4 | " | |
| " | " | " | " " | " | " | " | | 18.9 | " | 9.9 | 10.6 | " | |
| " | " | " | " 40.6 | " | " | " | | " | " | 10.2 | 10.9 | " | } Without side flue, with economizer. |
| " | " | " | " 50 | " | " | " | | 18.4 | " | 11.8 | 12.7 | " | " " |
| " | " | " | " 34.3 | " | " | " | | 21.8 | " | 10.8 | 11.6 | " | |
| " | " | " | " " | " | " | " | | 22.7 | " | 10.7 | 11.4 | " | |
| " | " | " | " " | " | " | " | | 22.6 | " | 11.1 | 12.0 | " | |
| " | " | " | " 74.7 | " | " | " | | 20.9 | " | 11.9 | 12.9 | " | With economizer. |
| " | 2.9 | 1.8 | 50.8 50 | " | 3.5 | 5.7 | | 27.6 | " | 11.4 | " | " | Average 15 trials. |
| " | " | " | " " | " | " | " | | 24.0 | " | 11.8 | " | " | |
| " | " | " | " " | " | " | " | | 25.0 | " | 12.4 | " | " | |
| " | 1.4 | 1.8 | " 27.9 | " | 3.5 | 5.7 | | 24.0 | " | 12.2 | " | " | Natural draft. |
| " | 2.9 | 1.8 | " 50 | " | 3.5 | 5.7 | | 27.6 | " | 11.5 | " | " | " " |
| " | " | " | " " | " | " | " | | 27.5 | " | 11.9 | " | " | |
| " | " | " | " " | " | " | " | | 41.2 | " | 11.4 | " | " | Fan draft. |
| " | " | " | " " | " | " | " | | 28.8 | " | 11.9 | " | " | |
| " | " | " | " " | " | " | " | | 26.2 | " | 12.4 | " | " | |
| " | " | " | " " | " | " | " | | 27.6 | " | 11.5 | " | " | |
| " | " | " | " " | " | " | " | | 25.5 | " | 12.5 | " | " | |
| " | " | " | " " | " | " | " | | 31.4 | " | 10.4 | " | " | |
| — | 5.7 | 4.9 | — 37.5 | — | 5.0 | 5.8 | | 21.1 | " | 8.9 | " | " | } Including heater; no air through door. |
| " | " | " | " " | " | " | " | | 19.0 | " | 11.1 | " | " | |
| " | " | " | " 55.6 | " | " | " | | 21.0 | " | 10.0 | " | " | |
| " | " | " | " " | " | " | " | | 17.2 | " | 12.5 | " | " | |
| " | " | " | " 48.6 | " | " | " | | 17.3 | " | 11.7 | " | " | |
| " | " | " | " " | " | " | " | | 27.0 | " | 10.8 | " | " | |
| " | " | " | " 59.6 | " | " | " | | 27.4 | " | 11.4 | " | " | |
| " | " | " | " 69 | " | " | " | | 37.4 | " | 10.6 | " | " | |
| " | " | " | " 25.5 | " | " | " | | 16.0 | " | 9.6 | " | " | |
| " | " | " | " " | " | " | " | | 17.6 | " | 9.1 | " | " | |
| " | " | " | " " | " | " | " | | 18.1 | " | 9.0 | " | " | |
| " | " | " | " 32.3 | " | " | " | | 20.4 | " | 9.2 | " | " | |
| " | " | " | " 34 | " | " | " | | 26.9 | " | 10.3 | " | " | Without heater. |

## STEAM MAKING; OR, BOILER PRACTICE.

| Number for Reference. | AUTHORITY. | LOCATION. | KIND OF BOILER. | KIND OF FUEL. | Grate area in square feet. | Water heating surface in sq. ft. |
|---|---|---|---|---|---|---|
| 186 | D. K. Clark | Newcastle | Return fire tube | Newcastle | 22 | 749 |
| 187 | " | " | " | " | " | " |
| 188 | " | " | " | " | " | " |
| 189 | " | " | " | " | Welsh | " | " |
| 190 | " | " | " | " | Newcastle | " | 1069 |
| 191 | " | " | " | " | " | " | " |
| 192 | " | " | " | " | " | " | " |
| 193 | " | " | " | " | " | " | " |
| 194 | " | " | " | " | " | " | " |
| 195 | " | " | " | " | " | " | " |
| 196 | " | " | " | " | " | " | " |
| 197 | " | " | " | " | " | " | " |
| 198 | " | " | " | " | " | " | " |
| 199 | " | " | " | " | " | " | " |
| 200 | " | " | " | " | " | " | " |
| 201 | " | " | " | " | " | " | " |
| 202 | " | " | " | " | Welch | " | " |
| 203 | " | " | " | " | " | " | " |
| 204 | " | " | " | " | " | " | " |
| 205 | " | " | " | " | " | " | " |
| 206 | " | " | " | " | " | " | " |
| 207 | " | " | " | " | " | " | " |
| 208 | " | " | " | " | " | " | " |
| 209 | " | " | " | " | Newcastle | 18 | " |
| 210 | " | " | " | " | " | " | " |
| 211 | " | Keyham Yard | " | English | 14 | 485 |
| 212 | " | " | " | " | " | " |
| 213 | " | " | " | " | " | " |
| 214 | " | " | " | " | " | " |
| 215 | " | " | " | " | " | " |
| 216 | " | " | " | " | " | " |
| 217 | " | " | " | " | " | " |
| 218 | " | " | " | " | " | " |
| 219 | " | " | " | " | " | " |
| 220 | " | " | " | " | " | " |
| 221 | " | " | " | " | " | " |
| 222 | " | " | " | " | " | " |
| 223 | " | " | " | " | " | " |
| 224 | " | " | " | " | " | " |
| 225 | " | " | " | " | " | " |
| 226 | " | " | " | " | " | " |
| 227 | " | " | " | " | " | " |
| 228 | " | " | " | " | " | " |
| 229 | " | " | " | " | " | 10.5 | " |
| 230 | " | " | " | " | " | " |
| 231 | " | " | " | " | " | " |
| 232 | " | " | " | " | " | " |
| 233 | " | " | " | " | " | " |
| 234 | R. H. Buel | N. Y. Navy Yard | " | Anthracite | 36 | 950 |
| 235 | " | " | " | " | " | " |
| 236 | " | " | " | " | " | " |
| 237 | " | " | " | " | " | " |
| 238 | " | " | " | " | " | " |
| 239 | " | " | " | " | " | " |
| 240 | " | " | " | " | " | " |
| 241 | " | " | " | " | " | 30 |
| 242 | " | " | " | " | " | 11 |

## ON COMBUSTION.

| Steam heating surface in sq. ft. | Least flue area in sq. ft. | Area of chimney. Sq. ft. | Height of chimney. ft. in. | Water heating surface ÷ grate area. | Steam heating surface ÷ grate area. | Grate area ÷ least flue area. | Grate area ÷ chimney area. | Lbs. coal per sq. ft. grate per hour. | Per cent. refuse. | EVAPORAT'N FROM & AT 212°. Lbs. water ⁄ lb. coal. | Lbs. water ⁄ lb. com. | Duration in hours. | REMARKS. |
|---|---|---|---|---|---|---|---|---|---|---|---|---|---|
| 5.7 | 4.9 | | | 34 | | | | 27.0 | | 10.0 | | | |
| " | " | | | " | | | | 35.6 | | 9.5 | | | |
| " | " | | | " | | | | 31.6 | | 8.2 | | | |
| " | " | | | | | | | 23.2 | | 10.1 | | | |
| " | " | | | 48.6 | | | | 22.1 | | 11.4 | | | |
| " | " | | | " | | | | 23.0 | | 10.6 | | | |
| " | " | | | " | | | | 26.0 | | 11.2 | | | |
| " | " | | | " | | | | 26.0 | | 10.3 | | | |
| " | " | | | " | | | | 28.5 | | 10.6 | | | |
| " | " | | | " | | | | 29.5 | | 9.6 | | | |
| " | " | | | " | | | | 17.3 | | 10.5 | | | |
| " | " | | | " | | | | 27.5 | | 9.9 | | | |
| " | " | | | " | | | | 23.4 | | 10.1 | | | |
| " | " | | | " | | | | 23.4 | | 10.8 | | | |
| " | " | | | " | | | | 24.0 | | 10.1 | | | |
| " | " | | | " | | | | 16.3 | | 12.0 | | | |
| " | " | | | " | | | | 18.3 | | 12.4 | | | |
| " | " | | | " | | | | 20.6 | | 12.6 | | | |
| " | " | | | " | | | | 20.9 | | 11.4 | | | |
| " | " | | | " | | | | 21.9 | | 11.6 | | | |
| " | " | | | " | | | | 22.4 | | 12.9 | | | |
| " | " | | | " | | | | 9.2 | | 10.9 | | | |
| " | " | | | " | | | | 24.1 | | 11.1 | | | |
| " | " | | | 59.6 | | | | 18.7 | | 11.1 | | | |
| " | " | | | " | | | | 24.9 | | 11.0 | | | |
| | | | | 35.3 | | | | 15.4 | | 10.4 | | | Common doors. |
| | | | | " | | | | 18.6 | | 9.2 | | | |
| | | | | " | | | | 15.4 | | 9.8 | | | |
| | | | | " | | | | 14.1 | | 10.1 | | | |
| | | | | " | | | | 15.7 | | 9.7 | | | |
| | | | | " | | | | 16.7 | | 11.0 | | | |
| | | | | " | | | | 18.3 | | 9.4 | | | |
| | | | | " | | | | 16.2 | | 10.6 | | | |
| | | | | " | | | | 15.4 | | 10.6 | | | |
| | | | | " | | | | 16.4 | | 9.4 | | | |
| | | | | " | | | | 17.5 | | 11.2 | | | |
| | | | | " | | | | 14.9 | | 10.8 | | | Perforated doors. |
| | | | | " | | | | 17.0 | | 9.6 | | | |
| | | | | " | | | | 17.4 | | 10.5 | | | |
| | | | | " | | | | 16.6 | | 10.6 | | | |
| | | | | " | | | | 17.4 | | 10.4 | | | |
| | | | | " | | | | 22.9 | | 9.3 | | | |
| | | | | " | | | | 18.4 | | 10.8 | | | |
| | | | | 47 | | | | 22.5 | | 11.3 | | | Common doors. |
| | | | | " | | | | 21.6 | | 11.1 | | | |
| | | | | " | | | | 21.8 | | 10.6 | | | |
| | | | | " | | | | 24.4 | | 11.4 | | | Perforated doors. |
| | | | | " | | | | 22.3 | | 11.6 | | | |
| 4.6 | 6.8 | 60.0 | | 26.4 | | 0.128 | | 10.4 | | | 9.6 | | |
| | | | | " | | | | 19.7 | | | 9.4 | | |
| | | | | " | | " | | 19.5 | | | 9.5 | | |
| | | | | " | | " | | 18.8 | | | 9.5 | | |
| | | | | " | | " | | 13.1 | | | 11.5 | | |
| | | | | " | | " | | 16.9 | | | 12.7 | | |
| | | | | " | | " | | 7.8 | | | 11.4 | | |
| | | | | 31.7 | | .153 | | 17.8 | | | 10.8 | | |
| | | | | " | | " | | 8.6 | | | 12.2 | | |

32        STEAM MAKING; OR, BOILER PRACTICE.

| Number for Reference. | AUTHORITY. | LOCATION. | KIND OF BOILER. | KIND OF FUEL. | Grate area in square feet. | Water heating surface in sq. ft. |
|---|---|---|---|---|---|---|
| 243 | R. H. Buel | N. Y. Navy Yard | Return fire tube | Anthracite | 24 | 950 |
| 244 | " | " | " | " | " | " |
| 245 | " | " | " | " | 18 | " |
| 246 | " | " | " | " | " | " |
| 247 | " | " | " | " | " | " |
| 248 | " | " | " | " | 13.5 | " |
| 249 | " | " | " | " | " | " |
| 250 | " | " | " | " | 3.6 | " |
| 251 | " | " | " | " | " | " |
| 252 | " | " | " | " | " | " |
| 253 | " | " | " | " | " | " |
| 254 | " | " | " | " | " | " |
| 255 | " | " | " | " | " | " |
| 256 | " | " | " | " | " | " |
| 257 | " | " | " | " | 28.8 | " |
| 258 | " | " | " | " | 21.6 | " |
| 259 | " | " | " | " | 29 | " |
| 260 | " | " | " | " | 30 | " |
| 261 | " | " | " | " | " | " |
| 262 | " | " | " | " | " | " |
| 263 | " | " | " | " | " | " |
| 264 | " | " | " | " | " | " |
| 265 | " | " | " | " | 22 | " |
| 266 | " | " | " | " | 36 | " |
| 267 | " | " | " | " | " | " |
| 268 | " | " | " | " | " | " |
| 269 | " | " | " | " | " | " |
| 270 | " | " | " | " | " | " |
| 271 | " | " | " | " | " | " |
| 272 | " | " | " | " | " | " |
| 273 | " | " | " | " | " | " |
| 274 | " | " | " | " | " | " |
| 275 | " | " | " | " | " | " |
| 276 | " | " | " | " | " | " |
| 277 | " | " | " | " | " | " |
| 278 | " | " | " | " | " | " |
| 279 | " | " | " | " | " | " |
| 280 | " | " | " | " | " | " |
| 281 | " | " | " | " | " | " |
| 282 | " | " | " | " | 36 | " |
| 283 | " | " | " | " | " | " |
| 284 | " | " | " | " | 27 | " |
| 285 | " | " | " | " | 36 | " |
| 286 | " | " | " | " | " | " |
| 287 | " | " | " | " | " | " |
| 288 | " | " | " | " | " | " |
| 289 | " | " | " | " | " | " |
| 290 | " | " | " | " | " | " |
| 291 | " | " | " | " | " | " |
| 292 | " | " | " | " | " | " |
| 293 | " | " | " | " | " | " |
| 294 | " | " | " | " | " | " |
| 295 | " | " | " | " | " | " |
| 296 | " | " | " | " | " | " |
| 297 | " | " | " | " | " | " |
| 298 | " | " | " | " | " | " |
| 299 | " | " | " | " | " | " |

## ON COMBUSTION.

| Steam heating surface in sq. ft. | Least flue area in sq. ft. | Area of chimney. Sq. ft. | Height of chimney. ft. in. | Water heating surface ÷ grate area. | Steam heating surface ÷ grate area. | Grate area ÷ least flue area. | Grate area ÷ chimney area. | Lbs. coal per sq. ft. grate per hour. | Per cent refuse. | EVAPORA-T'N FROM & AT 212°. Lbs. water ℔ lb. coal. | Lbs. water ℔ lb. com. | Duration in hours. | REMARKS. |
|---|---|---|---|---|---|---|---|---|---|---|---|---|---|
| 4.6 | 0.8 | 00 | 39.5 | | .192 | | | 17.7 | | 11.5 | | | |
| " | " | " | " | | " | | | 8.7 | | 11.6 | | | |
| " | " | " | 52.8 | | .256 | | | 19.1 | | 11.5 | | | |
| " | " | " | " | | " | | | 8.4 | | 12.3 | | | |
| " | " | " | 70.4 | | .341 | | | 10.0 | | 11.3 | | | |
| " | " | " | " | | " | | | 8.1 | | 12.3 | | | |
| " | " | " | 26.4 | | .128 | | | 7.1 | | 12.7 | | | |
| " | " | " | " | | " | | | 17.9 | | 10.0 | | | |
| " | " | " | " | | " | | | 3.7 | | 13.1 | | | |
| " | " | " | " | | " | | | 16.7 | | 10.0 | | | |
| " | " | " | " | | " | | | 12.1 | | 11.0 | | | |
| " | " | " | " | | " | | | 17.6 | | 10.4 | | | |
| " | " | " | " | | " | | | 15.7 | | 10.6 | | | |
| " | " | " | 33 | | .16 | | | 15.2 | | 11.5 | | | |
| " | " | " | 44 | | .213 | | | 18.5 | | 11.6 | | | |
| " | " | " | 32.8 | | .159 | | | 14.6 | | 11.6 | | | |
| " | " | " | 26.4 | | .128 | | | 18.4 | | 10.4 | | | |
| " | " | " | " | | " | | | 20.3 | | 7.5 | | | |
| " | " | " | " | | " | | | 18.6 | | 9.4 | | | |
| " | " | " | " | | " | | | 21.4 | | 7.4 | | | |
| " | " | " | " | | " | | | 18.6 | | 9 7 | | | |
| " | " | " | 43.2 | | .209 | | | 21.9 | | 11.2 | | | |
| " | " | " | 26.4 | | .106 | | | 18.6 | | 10.8 | | | |
| " | " | " | " | | .102 | | | 18.5 | | 10.8 | | | |
| " | " | " | " | | .095 | | | 18.6 | | 10.9 | | | |
| " | " | " | " | | .114 | | | 16.6 | | 11.4 | | | |
| " | " | " | " | | .1 | | | 16.7 | | 11.2 | | | |
| " | " | " | " | | .128 | | | 17.4 | | 11.3 | | | |
| " | " | " | " | | " | | | 18.6 | | 10.7 | | | |
| " | " | " | " | | " | | | 17.1 | | 10.5 | | | |
| " | " | " | " | | " | | | 18.7 | | 10.6 | | | |
| " | " | " | " | | " | | | 22.4 | | 10.0 | | | |
| " | " | " | " | | " | | | 21.7 | | 9.9 | | | |
| " | " | " | " | | " | | | 21.7 | | 9.2 | | | |
| " | " | " | " | | " | | | 21.6 | | 11.1 | | | |
| " | " | " | " | | " | | | 22.1 | | 11.4 | | | |
| " | " | " | " | | .114 | | | 15.2 | | 12 0 | | | |
| " | " | " | " | | .128 | | | 14 0 | | 12.7 | | | |
| " | " | " | 26.4 | | .128 | | | 15.2 | | 12.2 | | | |
| " | " | " | " | | " | | | 21.7 | | 11.8 | | | |
| " | " | " | 35.2 | | .148 | | | 19.0 | | 12.9 | | | |
| " | " | " | 26.4 | | .108 | | | 18.3 | | 11.2 | | | |
| " | " | " | 5.9 | | .067 | | | 18.7 | | 5.9 | | | |
| " | " | " | " | | " | | | 7.9 | | 8.2 | | | |
| " | " | " | " | | " | | | 13.7 | | 6.9 | | | |
| " | " | " | 26.4 | | .087 | | | 15.4 | | 11.5 | | | |
| " | " | " | " | | .111 | | | 15 5 | | 11.4 | | | |
| " | " | " | 21.1 | | .099 | | | 14.0 | | 11.6 | | | |
| " | " | " | 19.9 | | .085 | | | 13.4 | | 11.7 | | | |
| " | " | " | 17.7 | | .071 | | | 11.6 | | 12.8 | | | |
| " | " | " | 15.5 | | .057 | | | 11.2 | | 11.8 | | | |
| " | " | " | 13.4 | | .043 | | | 8.6 | | 12.3 | | | |
| " | " | " | 11.2 | | .028 | | | 5.8 | | 12.4 | | | |
| " | " | " | 9.1 | | .014 | | | 2.4 | | 13.6 | | | |
| " | " | " | 26.4 | | .083 | | | 14.1 | | 11.7 | | | |
| " | " | " | " | | " | | | 15.2 | | 11.5 | | | |
| " | " | " | " | | " | | | 14.4 | | 11.7 | | | |

# STEAM MAKING; OR, BOILER PRACTICE.

| Number for Reference. | AUTHORITY. | LOCATION. | KIND OF BOILER. | KIND OF FUEL. | Grate area in square feet. | Water heating surface in sq. ft. |
|---|---|---|---|---|---|---|
| 300 | R. H. Buel | N. Y. Navy Yard | Return fire tube | Anthracite | 36 | 950 |
| 301 | " | " | " | " | 31.2 | " |
| 302 | " | " | " | " | " | " |
| 303 | " | " | " | " | " | " |
| 304 | " | " | " | " | " | " |
| 305 | " | " | " | " | " | " |
| 306 | " | " | " | " | " | " |
| 307 | " | " | " | " | 36 | " |
| 308 | " | " | Return water tube | " | " | 1265 |
| 309 | " | " | " | " | 39 | " |
| 310 | " | " | " | " | " | " |
| 311 | " | " | " | " | " | " |
| 312 | " | " | " | " | 36 | " |
| 313 | " | " | " | " | 39 | " |
| 314 | " | " | " | " | 30 | " |
| 315 | " | " | " | " | " | " |
| 316 | " | " | " | " | 24 | " |
| 317 | " | " | " | " | " | " |
| 318 | " | " | " | " | 18 | " |
| 319 | " | " | " | " | " | " |
| 320 | " | " | " | " | " | " |
| 321 | " | " | " | " | 13.5 | " |
| 322 | " | " | " | " | 39 | " |
| 323 | " | " | " | " | " | " |
| 324 | " | " | " | " | " | " |
| 325 | " | " | " | " | 36 | " |
| 326 | " | " | " | " | " | " |
| 327 | " | " | " | " | 39 | " |
| 328 | " | " | " | " | 36 | " |
| 329 | " | " | " | " | 31.2 | " |
| 330 | " | " | " | " | 23.4 | " |
| 331 | " | " | " | " | 31.4 | " |
| 332 | " | " | " | " | 36 | " |
| 333 | " | " | " | " | " | " |
| 334 | " | " | " | " | " | " |
| 335 | " | " | " | " | " | " |
| 336 | " | " | " | " | " | " |
| 337 | " | " | " | " | 39 | " |
| 338 | " | " | " | " | 36 | " |
| 339 | " | " | " | " | 39 | " |
| 340 | " | " | " | " | " | " |
| 341 | " | " | " | " | 36 | " |
| 342 | " | " | " | " | " | " |
| 343 | " | " | " | " | " | " |
| 344 | " | " | " | " | " | " |
| 345 | " | " | " | " | 39 | " |
| 346 | " | " | " | " | 23.8 | " |
| 347 | " | " | " | " | 39 | " |
| 348 | " | " | " | " | " | " |
| 349 | " | " | " | " | " | " |
| 350 | " | " | " | " | " | " |
| 351 | " | " | " | " | " | " |
| 352 | " | " | " | " | " | " |
| 353 | " | " | " | " | " | " |
| 354 | " | " | " | " | " | " |
| 355 | " | " | " | " | " | " |
| 356 | " | " | " | " | " | " |

ON COMBUSTION.    35

| Steam heating surface in sq. ft. | Least flue area in sq. ft. | Area of chimney. Sq. ft. | Height of chimney. ft. in. | Water heating surface ÷ grate area. | Steam heating surface ÷ grate area. | Grate area ÷ least flue area. | Grate area ÷ chimney area. | Lbs. coal per sq. ft. grate per hour. | Per cent. refuse. | EVAPORAT'N FROM & AT 212°. Lbs. water ⅌ lb. coal. | Lbs. water ⅌ lb. com. | Duration in hours. | REMARKS. |
|---|---|---|---|---|---|---|---|---|---|---|---|---|---|
| 4.6 | 6.8 | 60 | | 40.4 | | .142 | | 11.4 | | | 11.6 | | |
| " | " | " | | " | | " | | 18.3 | | | 11.0 | | |
| " | " | " | | " | | " | | 14.5 | | | 11 5 | | |
| " | " | " | | " | | " | | 5.0 | | | 12.6 | | |
| " | " | " | | " | | " | | 3.4 | | | 11.2 | | |
| " | " | " | | 26.4 | | .100 | | 6.8 | | | 12 0 | | |
| " | " | " | | " | | " | | 10.3 | | | 11.8 | | |
| 5.54 | 6.78 | " | | 32.4 | | .142 | | 12.8 | | | 10.8 | | |
| " | " | " | | " | | " | | 12.3 | | | 12 1 | | |
| " | " | " | | " | | " | | 12.7 | | | 11.9 | | |
| " | " | " | | " | | " | | 12 5 | | | 11 8 | | |
| " | " | " | | " | | " | | 11.8 | | | 12.0 | | |
| " | " | " | | 35.1 | | .154 | | 11.5 | | | 12.3 | | |
| " | " | " | | 32.4 | | .142 | | 7.3 | | | 13.3 | | |
| " | " | " | | 42.2 | | .185 | | 11.6 | | | 13.0 | | |
| " | " | " | | " | | " | | 8.6 | | | 13.3 | | |
| " | " | " | | 52.7 | | .231 | | 12.2 | | | 13 5 | | |
| " | " | " | | " | | " | | 8.6 | | | 12.7 | | |
| " | " | " | | 70.3 | | .308 | | 12.7 | | | 13.3 | | |
| " | " | " | | " | | " | | 8.3 | | | 13.4 | | |
| " | " | " | | 93.7 | | .411 | | 12.0 | | | 12.8 | | |
| " | " | " | | " | | " | | 9.0 | | | 13.8 | | |
| " | " | " | | 32.4 | | .142 | | 6.5 | | | 13.5 | | |
| " | " | " | | " | | " | | 12.2 | | | 10.5 | | |
| " | " | " | | " | | " | | 3 4 | | | 13.7 | | |
| " | " | " | | 35.1 | | .154 | | 10.3 | | | 12.7 | | |
| " | " | " | | " | | " | | 11.9 | | | 12.2 | | |
| " | " | " | | 32.4 | | .142 | | 16 0 | | | 11.0 | | |
| " | " | " | | 35.1 | | .154 | | 15.1 | | | 11.0 | | |
| " | " | " | | 40.5 | | .178 | | 9.6 | | | 13.0 | | |
| " | " | " | | 54.1 | | .237 | | 12.2 | | | 13.2 | | |
| " | " | " | | 40.3 | | .177 | | 10.4 | | | 13.1 | | |
| " | " | " | | 35.1 | | .154 | | 18.4 | | | 10.6 | | |
| " | " | " | | " | | " | | 12.5 | | | 12.8 | | |
| " | " | " | | " | | " | | 14.7 | | | 13.0 | | |
| " | " | " | | " | | " | | 13.8 | | | 12 9 | | |
| " | " | " | | " | | " | | 14.7 | | | 12.9 | | |
| " | " | " | | 32.4 | | .142 | | 14.8 | | | 11.8 | | |
| " | " | " | | 35.1 | | .154 | | 17.4 | | | 11.6 | | |
| " | " | " | | 32.4 | | .142 | | 15.9 | | | 12.1 | | |
| " | " | " | | " | | " | | 16.1 | | | 12.0 | | |
| " | " | " | | 35.1 | | .154 | | 16.4 | | | 12.3 | | |
| " | " | " | | " | | .151 | | 14.5 | | | 12.0 | | |
| " | " | " | | " | | .154 | | 15.1 | | | 12.8 | | |
| " | " | " | | " | | " | | 16.3 | | | 12.6 | | |
| " | " | " | | 32.4 | | .142 | | 12.7 | | | 13.1 | | |
| " | " | " | | 53.1 | | .233 | | 15 9 | | | 12.7 | | |
| " | " | " | | 32.4 | | .129 | | 17.3 | | | 11.3 | | |
| " | " | " | | " | | .114 | | 17.0 | | | 11.5 | | |
| " | " | " | | " | | .142 | | 16 6 | | | 11.8 | | |
| " | " | " | | " | | .114 | | 17.1 | | | 11.4 | | |
| " | " | " | | " | | .125 | | 14.3 | | | 12.0 | | |
| " | " | " | | " | | .111 | | 12.9 | | | 12 4 | | |
| " | " | " | | " | | .101 | | 12.4 | | | 12.6 | | |
| " | " | " | | " | | .126 | | 15.4 | | | 11 2 | | |
| " | " | " | | 19.7 | | .11 | | 15.3 | | | 11.0 | | |
| " | " | " | | 17.0 | | .10 | | 9.5 | | | 13.3 | | |
| " | " | " | | 32.4 | | | | | | | | | |

36  STEAM MAKING; OR, BOILER PRACTICE.

| Number for Reference. | AUTHORITY. | LOCATION. | KIND OF BOILER. | KIND OF FUEL. | Grate area in square feet. | Water heating surface in sq. ft. |
|---|---|---|---|---|---|---|
| 357 | R. H. Buel | N. Y. Navy Yard | Return water tube | Anthracite | 30 | 1205 |
| 358 | " | " | " | " | " | " |
| 359 | " | " | " | " | 36 | " |
| 360 | " | " | " | " | 30 | " |
| 361 | " | " | " | " | " | " |
| 362 | " | " | " | " | " | " |
| 363 | " | " | " | " | " | " |
| 364 | " | " | " | " | " | " |
| 365 | " | " | " | " | 36 | " |
| 366 | " | " | " | " | 30 | " |
| 367 | " | " | " | " | " | " |
| 368 | " | " | " | " | " | " |
| 369 | " | " | " | " | " | " |
| 370 | " | " | " | " | " | " |
| 371 | " | " | " | " | " | " |
| 372 | " | " | " | " | 36 | " |
| 373 | " | " | " | " | 30 | " |
| 374 | " | " | " | " | " | " |
| 375 | " | " | " | " | " | " |
| 376 | " | " | " | " | " | " |
| 377 | " | " | " | " | " | " |
| 378 | " | " | " | " | " | " |
| 379 | " | " | " | " | " | " |
| 380 | " | " | " | " | " | " |
| 381 | " | " | " | " | " | " |
| 382 | " | " | " | " | " | " |
| 383 | " | " | " | " | " | " |
| 384 | " | " | Return fire tube | " | 18 | " |
| 485 | " | " | " | " | 30 | " |
| 386 | " | " | " | " | 18 | " |
| 387 | " | " | " | " | " | " |
| 388 | " | " | " | " | 30 | " |
| 389 | " | " | " | " | " | " |
| 390 | " | " | " | " | " | " |
| 391 | " | " | " | " | " | " |
| 392 | " | " | " | " | " | " |
| 393 | " | " | " | " | " | " |
| 394 | " | " | " | " | " | " |
| 395 | " | " | " | " | " | " |
| 396 | " | " | " | " | " | " |
| 397 | " | " | " | " | " | " |
| 398 | " | " | " | " | " | " |
| 399 | " | " | " | " | " | " |
| 400 | " | " | " | " | " | " |
| 401 | " | " | " | " | " | " |
| 402 | " | " | " | " | " | " |
| 403 | " | " | " | " | " | " |
| 404 | " | " | " | " | " | " |
| 405 | " | " | " | " | " | " |
| 406 | " | " | " | " | " | " |
| 407 | " | " | " | " | " | " |
| 408 | " | " | " | " | " | " |
| 409 | " | " | " | " | " | " |
| 410 | " | " | " | " | " | " |
|  | C. Lindé | Augsburg | Three French boilers each with two heaters | " | 21 | 644 |
|  | " | " | Two tubular return | " | 22 | 481 |
|  |  |  |  |  | 20.5 | 644 |
|  |  |  |  |  | 24 | 1302 |

## ON COMBUSTION.

| Steam heating surface in sq. ft. | Least flue area in sq. ft. | Area of chimney. Sq. ft. | Height of chimney. ft. in. | Water heating surface ÷ grate area. | Steam heating surface ÷ grate area. | Grate area ÷ least flue area. | Grate area ÷ chimney area. | Lbs. coal per sq. ft. grate per hour. | Per cent refuse. | EVAPORA-T'N FROM & AT 212°. Lbs. water ℔ lb. coal. | EVAPORA-T'N FROM & AT 212°. Lbs. water ℔ lb. com. | Duration in hours. | REMARKS. |
|---|---|---|---|---|---|---|---|---|---|---|---|---|---|
| 5.54 | 0.78 | | | 32.4 | | .10 | | 11.5 | | | 13.1 | | |
| " | " | | | " | | " | | 6.4 | | | 13.4 | | |
| " | " | | | 35.1 | | .154 | | 17.7 | | | 11.2 | | |
| " | " | | | 32.4 | | .142 | | 17.1 | | | 11.1 | | |
| " | " | | | " | | " | | 15.4 | | | 11.7 | | |
| " | " | | | " | | " | | 17.2 | | | 11.0 | | |
| " | " | | | " | | " | | 16.8 | | | 11.8 | | |
| " | " | | | " | | " | | 20.1 | | | 9.8 | | |
| " | " | | | 35.1 | | 154 | | 21.6 | | | 9.2 | | |
| " | " | | | 32.4 | | .142 | | 26.9 | | | 7.1 | | |
| " | " | | | " | | .1 | | 4.3 | | | 14.4 | | |
| " | " | | | " | | .142 | | 17.4 | | | 12.5 | | |
| " | " | | | " | | " | | 19.5 | | | 12.5 | | |
| " | " | | | " | | .1 | | 18.9 | | | 13.3 | | |
| " | " | | | " | | .111 | | 8.6 | | | 14.1 | | |
| " | " | | | 35.1 | | .154 | | 15.3 | | | 11.3 | | |
| " | " | | | 32.4 | | .100 | | 7.8 | | | 14.9 | | |
| " | " | | | " | | .111 | | 14.0 | | | 13.4 | | |
| " | " | | | " | | .110 | | 8.3 | | | 13.7 | | |
| " | " | | | " | | .142 | | 18.0 | | | 12.1 | | |
| " | " | | | " | | .100 | | 17.1 | | | 12.0 | | |
| " | " | | | " | | .142 | | 12.5 | | | 12.9 | | |
| " | " | | | 32.4 | | " | | 12.6 | | | 13.1 | | |
| " | " | | | 4.6 | | .068 | | 16.9 | | | 6.1 | | |
| " | " | | | " | | " | | 6.7 | | | 8.1 | | |
| 3.9 | " | | | " | | " | | 12.7 | | | 7.0 | | |
| " | " | | | 32.4 | 0.100 | " | | 17.5 | | | 12.7 | | |
| " | " | | | 32.4 | .0100 | " | " | 17.0 | " | " | 12.4 | | |
| " | " | | | 70.3 | .111 | " | " | 13.6 | " | " | 12.7 | | |
| " | " | | | 32.4 | .110 | " | " | 14.6 | " | " | 12.2 | | |
| " | " | | | 70.3 | .111 | " | " | 10.0 | " | " | 13.3 | | |
| " | " | | | " | " | " | " | 5.9 | " | " | 13.4 | | |
| " | " | | | 32.4 | .142 | " | " | 12.7 | " | " | 12.1 | | |
| " | " | | | " | " | " | " | 10.3 | " | " | 13.2 | | |
| " | " | | | " | " | " | " | 10.3 | " | " | 13.1 | | |
| " | " | | | " | " | " | " | 14.7 | " | " | 12.6 | | |
| " | " | | | " | " | " | " | 10.0 | " | " | 13.2 | | |
| " | " | | | " | " | " | " | 20.0 | " | " | 12.0 | | |
| " | " | | | 31.2 | " | " | " | 10.1 | " | " | 13.2 | | |
| " | " | | | " | " | " | " | 11.7 | | | 12.5 | | |
| " | " | | | " | .100 | " | " | 17.4 | " | " | 12.2 | | |
| " | " | | | 20.1 | .142 | " | " | 10.0 | " | " | 12.9 | | |
| " | " | | | " | " | " | " | 12.6 | " | " | 12.6 | | |
| " | " | | | 27.0 | " | " | " | 10.3 | " | " | 12.7 | | |
| " | " | | | " | " | " | " | 13.1 | " | " | 12.0 | | |
| " | " | | | 24.9 | " | " | " | 10.1 | " | " | 12.5 | | |
| " | " | | | " | " | " | " | 14.6 | " | " | 11.8 | | |
| " | " | | | 22.8 | " | " | " | 10.3 | " | " | 12.1 | | |
| " | " | | | " | " | " | " | 13.9 | " | " | 11.6 | | |
| " | " | | | 20.7 | " | " | " | 10.4 | " | " | 12.2 | | |
| " | " | | | " | " | " | " | 15.9 | " | " | 11.2 | | |
| " | " | | | 18.8 | " | " | " | 10.1 | " | " | 11.6 | | |
| " | " | | | " | " | " | " | 16.3 | " | " | 10.7 | | |
| " | " | | | 16.5 | " | " | " | 10.1 | " | 45 | 11.4 | | |

38　　　　STEAM MAKING; OR, BOILER PRACTICE.

| Number for Reference. | AUTHORITY. | LOCATION. | KIND OF BOILER. | KIND OF FUEL. | Grate area in square feet. | Water heating surface in sq. ft. |
|---|---|---|---|---|---|---|
|  | C. Lindé | Augsburg | One Lancashire with two cylinder heaters | Anthracite | 21 24.5 | 1302 1100 |
| 411 | " | " | " | Pensburg | 134 | 5473 |
| 412 | " | " | " | Saarbruck | 90 | " |
| 413 | C. E Emery | U. S. S. Rush | Return fire tube | Anthracite | 57 | 1573 |
| 414 | " | U. S. S. Dexter | " | " | 57 | " |
| 415 | " | U. S. S. Dallas | " | " | 57 | 1689 |
| 416 | " | U. S. S. Gallatin | North River | " | 55.2 | 1805 |
| 417 | " | " | " | " | " | " |
| 418 | " | Mulhouse | Lancashire | Ronchamp | 20.5 | 6125 |
| 419 | " | " | " | " | " | " |
| 420 | " | " | " | " | " | " |
| 421 | " | " | " | Saarbruck | " | " |
| 422 | " | " | Fairbairn | Ronchamp | " | 1017.5 |
| 423 | " | " | " | " | " | " |
| 424 | " | " | " | Saarbruck | " | " |
| 425 | " | " | French | Ronchamp | " | 607.6 |
| 426 | " | " | " | " | " | " |
| 427 | " | " | " | Saarbruck | " | " |
| 428 | R. H. Thurston | New York | Howard | Anthracite | 27 | 593 |
| 429 | " | " | " | " | " | " |
| 430 | L. E. Fletcher | " | Sinclair | Scotch | 39.5 | 1507 |
| 431 | " | " | " | Welch | " | " |
| 432 | " | " | " | Scotch | " | " |
| 433 | " | " | Lancashire | " | 36.6 | 698.5 |
| 434 | " | " | " | " | " | " |
| 435 | " | London, at South Metropolitan Gas Works | " | Gas Coke | 16 | 679 |
| 436 | " | " | " | Welch Coal | " | " |
| 437 | " | " | " | Breeze | 27.5 | " |
| 438 | " | " | " | Welch Coal | " | " |
| 439 | " | " | " | " | " | " |
| 440 | " | " | " | " | " | " |
| 441 | C. Lindé | Pfersee near Augsburg | Eight French | Saarbruck | 158 | 6659 |
| 442 | " | " | " | " | 130 | 6232 |
| 443 | M. Longridge | Oak Mill | Two Lancashire | Bituminous | 64.5 | 3030 |
| 444 | " | " | " | " | " | " |
| 445 | " | " | " | " | " | " |
| 446 | N. McDougall | " | Three Lancashire | " | 81.2 | 3820 |
| 447 | " | " | Two Lancashire | " | 64 | 3126 |
| 448 | E. A. Cowper | " | Three Cornish | " | 52.5 | " |
| 449 | M. Longridge | Blackburn | One Lancashire | " | 30.2 | 2135 |
| 450 | " | " | " | " | " | " |
| 451 | " | " | " | " | " | " |
| 452 | " | " | " | " | " | " |
| 453 | " | " | " | " | " | " |
| 454 | " | " | " | " | " | " |
| 455 | " | " | " | " | " | " |
| 456 | " | Pimlico for W. W. at Kimberly | One Composite | " | 10.5 | 644 |
| 457 | " | " | " | " | " | " |
| 458 | " | " | " | " | " | " |
| 459 | " | " | " | " | " | " |
| 460 | " | " | " | " | " | " |
| 461 | " | " | " | " | 15 | " |
| 462 | " | " | " | " | " | " |

## ON COMBUSTION.

| Steam heating surface in sq. ft. | Least flue area in sq. ft. | Area of chimney. Sq. ft. | Height of chimney. ft. in. | Water heating surface ÷ grate area. | Steam heating surface ÷ grate area. | Grate area ÷ least flue area. | Grate area ÷ chimney area. | Lbs. coal per sq. ft. grate per hour. | Per cent. refuse. | EVAPORAT'N FROM & AT 212°. Lbs. water ℔ lb. coal | Lbs. water ℔ lb. com. | Duration in hours. | REMARKS. |
|---|---|---|---|---|---|---|---|---|---|---|---|---|---|
| 3.9 | 26 | 120 | 40 | | | | | | 4.0 | 5.3 | | | |
| " | | | 55 | | | | | | 7.6 | 8.1 | | | |
| 7.7 | | | 27.6 | 7.4 | | | 11.4 | 21.1 | 8.6 | 10.8 | | | |
| " | | | 27.6 | " | | | 12.0 | 20.3 | 8.7 | 10.9 | | | |
| 7.8 | | | 29.6 | 7.3 | | | 13.3 | 20.5 | 8.7 | 11.0 | | | |
| 105 | 702 | | 32.6 | 2.0 | 7.7 | | 16.2 | 21.6 | 8.2 | 10.3 | | | |
| " | " | | " | | | | 15.3 | " | " | 10.6 | | | |
| " | " | | 29.8 | | | | 18.5 | 13.4 | 8.9 | 10.3 | | | |
| " | " | | " | | | | 10.1 | 14.6 | 9.3 | 10.0 | | | |
| " | " | | " | | | | 19.0 | 14.4 | 7.8 | 8.7 | | | |
| " | " | | " | | | | 16.1 | 9.7 | 9.0 | 10.6 | | | |
| " | " | | 49.5 | | | | 18.0 | 13.8 | 9.6 | 11.4 | | | |
| " | " | | " | | | | 10.4 | 13.4 | 9.0 | 10.4 | | | |
| " | " | | " | | | | 16.1 | 10.5 | 8.3 | 9.2 | | | |
| " | " | | 30.3 | | | | 20.1 | 14.5 | 8.6 | 10.0 | | | |
| " | " | | " | | | | 11.1 | 13.6 | 9.6 | 10.4 | | | |
| " | " | | " | | | | 16.9 | 9.1 | 7.9 | 8.7 | | | |
| 294 | 1.9 | 4.4 | 60 | 22 | 10.9 | 14.1 | 6.2 | comb. 6.8 coal. | | " " | 10.7 10.3 | | |
| | | | | 38.1 | | | | 13.2 | | 10.4 | | | |
| | | | | " | | | | 10.7 | | 12.2 | | | |
| | | | | " | | | | 16.6 | | 8.3 | | | |
| | | | | 29 | | | | 23.2 | | 6.6 | | | |
| | | | | " | | | | 22.8 | | 6.5 | | | |
| | | | | 42.5 | | | | 8.4 | | 7.5 | | | |
| | | | | " | | | | 8.1 | | 11.4 | | | |
| | | | | 24.7 | | | | 4.9 | | 7.4 | | | |
| | | | | " | | | | 4.1 | | 9.5 | | | |
| | | | | " | | | | 5.3 | | 8.8 | | | |
| | | | | " | | | | 6.4 | | 8.0 | | | |
| | | 46 | 161 | 42 | | | | | | 4.6 | | | |
| | | | | 44.4 | | | | | | 4.7 | | | |
| | | | | 47 | | | | 16.4 | | 9.6 | 11.6 | | |
| | | | | | | | | 15.6 | | 10.0 | 12.0 | | |
| | | | | | | | | 15.6 | | 10.8 | 12.7 | | |
| | | | | 47 | | | | 22.4 | 11.0 | 7.4 | 8.3 | | |
| | | | | 49 | | | | 19.8 | 8.2 | 9.8 | 11.0 | | |
| | | | | " | | | | | | 8.3 | | | |
| | | | | 71 | | | | 19.6 | 11 | 10.2 | 11.4 | | |
| | | | | | | | | 20.0 | 10 | 10.4 | 11.5 | | |
| | | | | | | | | 20.0 | 9 | 10.4 | 11.5 | | |
| | | | | | | | | 19.0 | | 10.4 | | | |
| | | | | | | | | 19.3 | 10 | 9.6 | 11.1 | | |
| | | | | | | | | 20.8 | 8 | 9.9 | 10.8 | | |
| | | | | | | | | 22.6 | 9 | 9.5 | 10.7 | | |
| | | | | 11.3 | | | | 11.4 | | 10.5 | | | |
| | | | | " | | | | 8.5 | | 10.8 | | | |
| | | | | " | | | | 5.7 | | 10.8 | | | |
| | | | | " | | | | 4.2 | | 8.1 | | | |
| | | | | 43 | | | | 8.0 | | 11.2 | | | |
| | | | | | | | | 6.7 | | 11.5 | | | |
| | | | | | | | | 5.4 | | 11.5 | | | |

| Number for Reference | AUTHORITY | LOCATION | KIND OF BOILER | KIND OF FUEL | Grate area in square feet | Water heating surface in sq. ft. |
|---|---|---|---|---|---|---|
| 463 | M. Longridge | Pimlico for W.W. at Kim'ly. | One composite | Bituminous | 15 | 644 |
| 464 | " | " | " | " | " | " |
| 465 | " | " | " | " | 22 | " |
| 466 | " | " | " | " | " | " |
| 467 | " | " | " | " | " | " |
| 468 | Report Committee | Dusseldorf | Lancashire, with tubular boiler | Essen coal | 33.4 | 1520 |
| 469 | " | " | Heine water tube | " | 19.4 | 869 |
| 470 | " | " | Cornish, with ret'n tubes | " | 19.7 | 1194 |
| 471 | " | " | Lancashire, with Galloway tubes | " | 36.6 | 1068 |
| 472 | " | " | Cornish | " | 15.8 | 1034 |
| 473 | " | " | Steinmuller's water tube | " | 32.3 | 724 |
| 474 | " | " | Neuman's, with water pockets | " | 13.7 | 494 |
| 475 | " | " | Lancashire, with Galloway tubes, and tubular boiler also | " | 34.8 | 2143 |
| 476 | " | " | Cornish | " | 15.8 | 389 |
| 477 | " | " | Lancashire, with Galloway tubes, and tubular above | " | 34.8 | 2142 |
| 478 | " | " | Büttner's water tube | " | 12.6 | 1942 |
| 479 | " | " | Watther's water tube | " | 15.7 | 996 |
| 480 | J. W. Hill | Evansville W. W. | 2 ext. fired ret. 6" tube | Anthracite | 45 | 329 |
| 481 | " | " | " | " | | |
| 482 | " | Cincinnati W. W. | 2 ext. fired ret. flue | Pittsburgh | 19 | 1083 |
| 483 | " | " | " | " | | |
| 484 | " | Holly at Buffalo, N. Y. | 2 North River | " | 42.7 | 1977 |
| 485 | R. H. Buel | Lawrence W. W. | Locomotive | Cumberland | 28.75 | 1020 |
| 486 | Report Committee | " | " | " | | |
| 487 | " | " | " | " | | |
| 488 | " | Brooklyn W. W. | Return drop flue | Anthracite | 38 | 818 |
| 489 | " | Chicago N. W. W. | Marine | " | | |
| 490 | C. Hermany | Cincinnati W. W. | Flue ext. fired | Pittsburgh | 14.7 | |
| 491 | " | " | Flue int. fired | " | 34.3 | |
| 492 | " | " | Flue ext. fired | " | | |
| 493 | " | " | " | " | | |
| 494 | " | " | " | " | | |
| 495 | " | " | " | " | | |
| 496 | " | " | Flue int. fired | " | 20.7 | |
| 497 | " | Hartford W. W. | " | Anthracite | 23.1 | 1122 |
| 498 | " | " | " | " | | |
| 499 | " | " | " | " | | |
| 500 | " | Jersey City W. W. | " | " | 20.2 | 817 |
| 501 | " | Louisville W. W. | Return drop flue | Pittsburgh | 22.5 | 675 |
| 502 | " | Lowell W. W. | Return tube ext. fired | Anthracite | 27.5 | 1100 |
| 503 | " | Lynn W. W. | " | " | 32 | 818 |
| 504 | Johnson | Washington W. W. | Return flue ext. fired | " | 14.3 | 377 |
| 505 | " | " | " | Soft | 14.1 | |
| 506 | " | " | " | Coking | " | " |
| 507 | P. W. Schaumleffel | St. Louis Lead & Oil Co. | " | " | Illinois | 51 | 1224 |

## ON COMBUSTION.

| Steam heating surface in sq. ft. | Least flue area in sq. ft. | Area of chimney. Sq. ft. | Height of chimney. ft. in. | Water heating surface ÷ grate area. | Steam heating surface ÷ grate area. | Grate area ÷ least flue area. | Grate area ÷ chimney area. | Lbs. coal per sq. ft. grate per hour. | Per cent refuse. | Evapor'n from & at 212°. Lbs. water ℔ lb. coal. | Lbs. water ℔ lb. com. | Duration in hours. | REMARKS. |
|---|---|---|---|---|---|---|---|---|---|---|---|---|---|
| | | | | | | | | 4.0 | | 11.0 | | | |
| | | | | | | | | 2.7 | | 8.3 | | | |
| | | | 9.2 | | | | | 8.0 | | 10.8 | | | |
| | | | | | | | | 3.6 | | 11.2 | | | |
| | | | | | | | | 6.8 | | 11.2 | | | |
| 5.6 | | | | | | 43 | | 11.2 | | 10.9 | 11.6 | | |
| 5.3 | | | | | | 45 | | 18.8 | | 9.0 | 10.8 | | |
| 2.3 | | | | | | 50 | | 19.7 | | 11.0 | 12.0 | | |
| 4.7 | | | | | | 29 | | 11.4 | | 10.0 | 11.1 | | |
| 4.4 | | | | | | 53 | | 20.4 | | 11.5 | 12.2 | | |
| 4.5 | | | | | | 22 | | 13.8 | | 8.8 | 10.6 | | |
| 4.0 | | | | | | 36 | | 14.5 | | 8.7 | 10.6 | | |
| 5.3 | | | | | | 24 | | 14.6 | | 11.1 | 12.0 | | |
| 4.2 | | | | | | 55 | | 19.5 | | 8.5 | 9.3 | | |
| 5.3 | | | | | | 54 | | 15.6 | | 11.0 | 12.2 | | |
| 6.4 | | | | | | 74 | | 31.0 | | 10.5 | 11.4 | | |
| 3.4 | | | | | | 41 | | 17.5 | | 8.8 | 9.9 | | |
| 4.7 | | | | | | 2.70 | 9.6 | 8.0  18.6 | | 8.4 | 10.3 | | |
| | | | | | | " | | 8.2  11.5 | | 8.2 | 9.3 | | |
| 2.5 | | | | | | 56.9 | 7.6 | 20.4  3 | | 10.6 | 10.9 | | |
| | | | | | | | | 21.0 | | 10.0 | 10.3 | | |
| 5.8 | | 76 | | | | 46.1 | .139 | 6.45  6.39 | | 13.6 | 14.5 | | |
| | | 120 | | | | 35.5 | .117 | 8.36  6.14 | | 11.5 | 12.2 | | |
| | | | | | | | | 7.07 | | } 9.3 | | | |
| | | | | | | | | 9.22 | | | | | |
| | | | | | | 21.5 | | | | 11.1 | | | |
| | | | | | | | | 13.1 | | 11.1 | | | |
| | | | | | | | | 10.2 | | 8.5 | | | |
| | | | | | | | | 20.4 | | 8.5 | | | |
| | | | | | | | | 12.3 | | 8.4 | | | |
| | | | | | | | | 16.1 | | 8.5 | | | |
| | | | | | | | | 12.3 | | 8.3 | | | |
| | | | | | | | | 22.8 | | 8.8 | | | |
| | | | | | | | | | | 9.5 | | | |
| | | | | | | 48.5 | | | | | | | |
| | | | | | | | | 6.9 | | 12.4 | | | |
| | | | | | | | | 5.6 | | 11.5 | | | |
| | | | | | | | | — | | 12.5 | | | |
| | | | | | | 40.4 | | 4.9 | | 11.5 | | | Mean of 4. |
| | | | | | | 30 | | | | 10.1 | | | |
| | | | | | | 40 | | | | 9.9 | | | |
| | | | | | | 25.6 | | 13.9 | | 11.0 | | | |
| | | | | | | 26.8 | | 6.6 | | 9.6 | | | 7 samples. |
| | | | | | | 27.0 | | 7.0 | | 9.7 | | | 11 samples. |
| | | | | | | " | | 7.4 | | 8.5 | | | 10 samples. |
| | | | | | | 24 | | 13.3 | | 5.44 | | 1 week. | |

42        STEAM MAKING; OR, BOILER PRACTICE.

| Number for Reference. | AUTHORITY. | LOCATION. | KIND OF BOILER. | KIND OF FUEL. | Grate area in square feet. | Water heating surface in sq. ft. |
|---|---|---|---|---|---|---|
| 508 | P. W. Schaumleffel | St. Louis Lead & Oil Co. | Return flue ext. fired | Illinois | 51 | 1224 |
| 509 | " | " | " " " | " | " | " |
| 510 | " | " | " " " | " | " | " |
| 511 | " | " | " " " | " | " | " |
| 512 | " | " | " " " | " | " | " |
| 513 | C. A. Smith | St. Louis, W'n University | Return tube | " | 24 | 768 |
| 514 | " | " | " " | " | 48 | 1536 |
| 515 | " | " | " " | " | " | " |
| 516 | B. F. Isherwood and Com'n | S. S. Leila | Herreshof coil | Anthracite | 26 | 485 |
| 517 | " | " | " " | " | " |  |
| 518 | " | " | " " | " | " |  |
| 519 | " | " | " " | " | " |  |
| 520 | " | " | " " | " | " |  |
| 521 | " | " | " " | " | " |  |
| 522 | " | " | " " | " | " |  |
| 523 | " | " | " " | " | " |  |
| 524 | C. H. Loring and Commission | S S. Anthracite | Perkins | Cumberland | 15.3 | 300 |
| 525 | C. A. Smith | St. Louis | Heine | Illinois | 8.3 | 272 |
| 526 | " | " | " | " | " |  |
| 527 | " | " | " | " | " |  |
| 528 | " | " | Herreshof | Anthracite | 20.3 | 330 |
| 529 | F. H. Pond | Edwardsville | Flue return ext. fired | Mt. Olive | " |  |
| 530 | " | " | " " " " | " |  |  |
| 531 | " | " | " " " " | Edwardsville | 36 | 932 |
| 532 | J. W. Hill | Saratoga | Return tubular ext. fired | " | 57 | 21515 |
| 533 | " | Newport, Ky | Return flue ext. fired | " | 35 | 872 |
| 534 | " | " | " " " " | " |  |  |

ON COMBUSTION.    43

| Steam heating surface in sq. ft. | Least flue area in sq. ft. | Area of chimney. Sq. ft. | Height of chimney. ft. in. | Water heating surface ÷ grate area. | Steam heating surface ÷ grate area. | Grate area ÷ least flue area. | Grate area ÷ chimney area. | Lbs. coal per sq. ft. grate per hour. | Per cent. refuse. | EVAPORAT'N FROM & AT 212°. Lbs. water ℔ coal | Lbs. water ℔ com. | Duration in hours. | REMARKS. |
|---|---|---|---|---|---|---|---|---|---|---|---|---|---|
|  |  |  |  | 24 |  |  |  | 26 |  | 6.30 |  | 1 week. |  |
|  |  |  |  | " |  |  |  |  |  | 6.54 |  | 72 hours |  |
|  |  |  |  | " |  |  |  |  |  | 5.70 |  | 2 weeks. |  |
|  |  |  |  | " |  |  |  |  |  | 6.5 |  | 2 weeks. |  |
|  |  |  |  | " |  |  |  |  |  | 7.1 |  | 3 weeks. |  |
|  | 3.14 | 12.25 | 100 | 32 |  | 7.6 | 2.0 | 38 | 12.5 | 6.9 | 8.6 | 1 week. | Coal has heating power .80 of carbon. |
|  | 6.28 |  |  | " |  |  | 4.0 | 24 |  | 7.1 | 9.3 | 1 week. |  |
|  |  |  |  | " |  |  | " |  |  | 7.9 | 9.9 | 24 hours |  |
| 44 | 10.0 | 3.14 | 27 | 18.7 | 1.7 | 3.9 | 8.26 | 12.8 | 15 | 8.2 | 9.7 |  |  |
|  |  |  |  |  |  |  |  | 11.8 | 18 | 8.1 | 9.9 |  |  |
|  |  |  |  |  |  |  |  | 11.3 | 16 | 8.2 | 9.8 |  |  |
|  |  |  |  |  |  |  |  | 8.3 | 20 | 9.1 | 11.5 |  |  |
|  |  |  |  |  |  |  |  | 7.9 | 21 | 8.4 | 10.6 |  |  |
|  |  |  |  |  |  |  |  | 5.0 | 20 | 8.3 | 10.4 |  | -- |
|  |  |  |  |  |  |  |  | 4.9 | 22 | 8.6 | 10.9 |  |  |
|  |  |  |  |  |  |  |  | 8.1 | 13 | 9.1 | 11.2 |  |  |
| 320 | 3.7 |  |  | 29.8 | 30 | 4.3 |  | 12.0 | 17.6 | 9.3 | 11.3 |  |  |
|  |  | 1.8 | 56 | 32.7 |  |  | 4.5 | 22.5 | 14 | 6.0 | 6.9 |  |  |
|  |  |  |  |  |  |  |  | 32 | 10 | 5.6 | 6.2 |  |  |
|  |  |  |  |  |  |  |  | 40 | 8 | 5.8 | 6.3 |  |  |
|  |  |  | 125 | 16.2 |  |  |  | 10.3 | 16 | 7.3 | 8.8 |  |  |
|  |  |  |  | 26 |  |  |  | 16.6 | 12.8 | 5.9 | 6.8 |  |  |
|  |  |  |  |  |  |  |  | 17.9 | 11.5 | 6.0 | 6.7 |  |  |
|  |  |  |  |  |  |  |  | 14.8 | 21.5 | 7.2 | 8.4 |  |  |
|  |  |  |  | 31.9 |  |  |  | 5.8 | 3.2 | 11.3 | 11.6 |  | Bearing's setting. |
|  | 5.0 | 11.1 | 60 | 24.9 |  |  | 6.26 | 15.3 | 9.8 | 8.5 | 9.4 |  | Common. |
|  |  |  |  |  |  |  |  | 25.9 | 4.4 | 6.2 | 6.5 |  |  |

## LOCOMOTIVE BOILERS.

| Number for Reference. | NAME. | Area of grate. Sq. ft. | Area of heating surface. Sq. ft. | Heating surface ÷ grate surface. | Coke per sq. ft. of grate per hr. in lbs. | Water per sq. ft. of grate per hour in cubic ft. | Pounds water evaporated per lb. fuel from and at 212° | AUTHORITY. |
|---|---|---|---|---|---|---|---|---|
| | Earliest Engs. | | | | coal | | | |
| 1 | Killingworth... | 7.0 | 41 | 6 | 44 | 2.3 | 4.02 | D. K. Clark. |
| 2 | " | 10.9 | 124 | 11.4 | 57 | 4 | 5.32 | " |
| | | | | | coke | | | |
| 3 | Rocket......... | 6 | 138 | 23 | 35.5 | 3 | 6.27 | " |
| 4 | Phoenix........ | 6 | 326 | 55 | 54 | 5.7 | 7.86 | " |
| 5 | Atlas.......... | 9.2 | 275 | 30 | 60 | 5.14 | 6.35 | " |
| 6 | Star........... | 7.76 | 359 | 46 | 92 | 8.22 | 6.53 | " |
| 7 | Average of 4... | 6.5 | 348 | 53.5 | 90 | 9.8 | 8.04 | " |
| 8 | ⎱ | | | | 100 | 10 | 7.42 | " |
| 9 | ⎰ Soho........ | 8.44 | 412 | 35 | 130 | 13.03 | 7.38 | " |
| 10 | | | | | 92 | 11.0 | 8.87 | " |
| 11 | Hecla.......... | 8.34 | 418 | 49 | 125 | 11.3 | 6.65 | " |
| 12 | Bury'o......... | 9.2 | 461 | 50 | 111 | 9.24 | 6.15 | " |
| 13 | " | 9.2 | 387 | 42 | 112 | 8.15 | 4.93 | " |
| | G. W. R. | | | | | | | |
| 14 | Ixion.......... | 13.4 | 699 | 52 | 138 | 15 | 8.33 | " |
| 15 | Hercules....... | 13.6 | 699 | 51.4 | 105 | 15 | 10.70 | " |
| 16 | Etna & 1...... | 11.4 | 467 | 41. | 97 | 10.7 | 8.21 | " |
| 17 | Giraffe........ | 12.5 | 608 | 48.6 | 76 | 8.8 | 8.61 | " |
| 18 | Mentor & 1.... | 13.6 | 699 | 51.4 | 69 | 8.0 | 8.67 | " |
| 19 | Royal Star..... | 11.7 | 822 | 70.0 | 91 | 10.8 | 8.85 | " |
| 20 | Pyracmon Cl.. | 18.44 | 1363 | 74.0 | 69 | 8.4 | 9.09 | " |
| 21 | Ajax........... | 13.67 | 1067 | 78 | 84 | 11.2 | 9.90 | " |
| 22 | Great Britain | 21 | 1938 | 92 | 82 | 11 | 9.95 | " |
| 23 | Great Britain class... | 21 | 1938 | 92 | 90 | 11 | 9.17 | " |
| 24 | Courier class... | 23.62 | 1866 | 79 | 75 | 8.6 | 8.60 | " |
| | L. & N. W. R. | | | | | | | |
| 25 | A.............. | 9.6 | 903 | 94 | 132 | 17 | 10.52 | " |
| 26 | Hercules....... | 9.6 | 828 | 86 | 105 | 15 | 10.7 | " |
| 27 | Sphinx......... | 19.56 | 1056 | 100 | 157 | 22.1 | 10.41 | " |
| 28 | Heron.......... | 10.5 | 782 | 74.5 | 90 | 11.1 | 9.20 | " |
| 29 | No. 291........ | 19 | 1449 | 76.26 | 56.5 | 6.2 | 8.23 | " |
| 30 | No. 300........ | 22 | 1263 | 57.41 | 50.7 | 6.6 | 9.28 | " |
| | S. E. R. | | | | coal | | | |
| 31 | No. 142........ | 14.7 | 1158 | 78.8 | 62.25 | | 10.15 | " |
| 32 | No. 118........ | 26.25 | 963 | 36.7 | 38.86 | | 10.60 | " |
| 33 | No. 58......... | 12.25 | 706 | 57.6 | 61.22 | | 10.13 | " |
| 34 | No. 58......... | 12.25 | 706 | 57.6 | 44.49 | | 11.91 | " |
| | | | | | coke | | | |
| 35 | No. 142........ | 14.7 | 1158 | 78.8 | 51.71 | | 9.77 | " |
| 36 | No. 105........ | 10.5 | 623 | 59.3 | 55.91 | | 11.68 | " |
| 37 | No. 9.......... | 10.5 | 623 | 59.3 | 66.19 | | 10.96 | " |
| | L. & S. W. R. | | | | | | | |
| 38 | Snake.......... | 12.4 | 985 | 79 | 87 | 12.26 | 10.59 | " |
| | | | | | coal | | | |
| 39 | Canute........ | 16 | 871 | 54.4 | 35 | 6.18 | 11.02 | " |
| 40 | " | " | " | " | 57 | 9.65 | 10.57 | " |
| 41 | " | " | " | " | 49 | 7.77 | 9.90 | " |

ON COMBUSTION. 45

| Number for Reference. | NAME. | Area of grate. Sq. ft. | Area of heating surface. Sq. ft. | Heating surface ÷ grate surface. | Coke per sq. ft. of grate per hr. in lbs. | Water per sq. ft. of grate per hour in cubic ft. | Pounds water evaporated per lb. fuel from and at 212° | | AUTHORITY. |
|---|---|---|---|---|---|---|---|---|---|
| 42 | Canute........ | 16 | 871 | 54.4 | 42 | 6.42 | 9.54 | | D. K. Clark. |
| 43 | " ........ | " | " | " | 58 | 8.89 | 9.56 | | " |
| | | | | | coke | | | | |
| 44 | " ........ | " | " | " | 46 | 6.46 | 8.76 | | " |
| 45 | " ........ | " | " | " | 49 | 7.19 | 9.13 | | " |
| 46 | " ........ | " | " | " | 54 | 8.69 | 10.04 | | " |
| | Caledonian R. | | | | | | | | |
| 47 | No. 33......... | 10.5 | 831 | 79 | 42 | 7 | 12.46 | | " |
| 48 | No. 42......... | " | 788 | 75 | 57 | 7.8 | 10.11 | | " |
| 49 | No. 43......... | " | " | 75 | 61 | 9.2 | 11.31 | | " |
| 50 | No. 51......... | " | " | 75 | 45 | 6.7 | 11.04 | | " |
| 51 | No. 13......... | " | " | 75 | 108 | 11.6 | 8.09 | | " |
| 52 | No. 13......... | " | " | 75 | 57 | 8.2 | 10.71 | | " |
| 53 | No. 13......... | 9.0 | " | 87.6 | 102 | 14.1 | 9.52 | | " |
| 54 | No. 125, 127... | 11.37 | 1050 | 92 | 66 | 18.1 | 9.72 | | " |
| 55 | No. 102......... | 11.8 | 974 | 82.5 | 94 | 10.3 | 8.15 | | " |
| | E. & G. R. | | | | | | | | |
| 56 | Orion & Sirius | 12.23 | 758 | 62 | 44 | 6.3 | 10.71 | | " |
| 57 | America & Nile | 11.10 | 736 | 66.3 | 70 | 6.8 | 9.32 | | " |
| 58 | Pallas........... | 16.04 | 818 | 51 | 38 | 6.0 | 10.47 | | " |
| 59 | Brindley....... | 9.15 | 802 | 87.65 | 54 | 7.2 | 9.94 | | " |
| | G. & S W. R. | | | | | | | | |
| 60 | Orion.......... | 9.24 | 495 | 53.6 | 84 | 9.4 | 8.29 | | " |
| 61 | Queen......... | 10.5 | 688 | 65.5 | 87 | 10.0 | 8.57 | | " |
| | P. F. W. & C.R. | | | | coal | | | | |
| 62 | No. 121........ | 15 | 831 | 55.4 | 62 | | 7.88 | | Reports M. M. |
| 63 | No. 158........ | 18 | 1050 | 58.3 | 109 | | 6.6 | | Assoc. U. S. |
| 64 | | | | | 119 | | 6.18 | | D. K. Clark. |
| 65 | | | | | 163 | | 6.07 | | " |
| 66 | | | | | 119 | | 7.05 | | " |
| 67 | | | | | 118 | | 6.46 | | " |
| 68 | | | | | 111 | | 7.28 | | " |
| | Jeff. Mad. & Ind. | | | | | | | | |
| 69 | | 10 | 779 | 77.9 | 131.6 | | 5.70 | | " |
| 70 | | " | " | " | 90.5 | | 8.34 | | " |
| | | | | | coal. | | | | |
| 71 | | " | " | " | 82.2 | | 6.66 | | " |
| 72 | | 11 | 708 | 64.3 | 69.6 | | 6.68 | | " |
| 73 | | " | " | " | 56.6 | | 6.61 | | " |
| 74 | No. 28......... | 13 | 920 | 70.8 | 74.7 | | 7.54 | | " |
| 75 | No. 28......... | 13 | 720 | 55.4 | 77.0 | | 5.64 | | " |
| 76 | No. 7........... | 12.5 | 707 | 56.6 | 102.5 | | 6.65 | | " |
| 77 | No. 30......... | 12 | 694 | 57.8 | 80.7 | | 7.28 | | " |
| | B. & A. R. R. | | | | | | Coal. | Comb. | |
| 78 | No. 129......... | 17.4 | 1245 | 71.5 | 121.5 | | 6.17 | | R. R. Gazette. |
| 79 | " ......... | | | | 122.2 | | 6.36 | 7.01 | " |
| 80 | " ......... | | | | 94.8 | | 7.31 | 7.92 | " |
| 81 | " ......... | | | | 91.0 | | 7.61 | 8.27 | " |
| 82 | No. 169......... | " | " | " | 115.5 | | 6.09 | 7.48 | " |
| 83 | " ......... | | | | 106.4 | | 7.08 | 7.85 | " |
| 84 | " ......... | | | | 93.3 | | 7.53 | 8.17 | " |

# STEAM MAKING; OR BOILER PRACTICE.

| Number for Reference. | NAME. | Area of grate. Sq. ft. | Area of heating surface. Sq. ft. | Heating surface ÷ grate surface. | Coke per sq. ft. of grate per hr. in lbs. | Water per sq. ft. of grate per hour in cubic ft. | Pounds water evaporated per lb. fuel from and at 212° | | AUTHORITY. |
|---|---|---|---|---|---|---|---|---|---|
| | | | | | | | Coal. | Comb. | |
| 85 | No. 160......... | 17.4 | 1245 | 71.5 | 86.3 | 10.0 | 8.01 | 8.64 | R. R. Gazette. |
| 86 | No. 150......... | 16 | 1012 | 63.3 | 130.0 | | 6.42 | 6.97 | " |
| 87 | " .......... | | | | 114.9 | | 6.67 | 7.25 | " |
| | C. H. & D. | | | | | | | | |
| 88 | No. 36........ | 15.1 | 899 | 59.6 | 83.9 | | 8.36 | | M. M. Ass'n. |
| 89 | " .......... | | | | 171.8 | | 5.34 | | " |
| 90 | " .......... | | | | 117.3 | | 7.30 | | " |
| | W. St. L. & P. | | | | | | | | |
| 91 | No. 180......... | 15.3 | 848 | 55.5 | 109 | | 5.86 | 7.32 | " |
| 92 | | | | | 118 | | 5.08 | 7.52 | " |
| 93 | | | | | 115 | | 5.98 | 7.50 | " |
| 94 | | | | | 37 | | 8.48 | 10.60 | " |
| 95 | | | | | 105 | | 7.21 | 9.00 | " |
| 96 | | | | | 111 | | 6.48 | 8.1 | " |
| 97 | | | | | 113 | | 6.60 | 8.2 | " |
| 98 | | | | | 92 | | 7.44 | 9.3 | " |
| 99 | | | | | 96 | | 6.10 | 7.7 | " |
| 100 | No. 144 ... .... | 13.1 | 846 | 64.6 | 125 | | 5.78 | 7.2 | " |
| 101 | | | | | 200 | | 5.3 | 6.6 | " |
| 102 | | | | | 110 | | 6.06 | 8.4 | " |
| 103 | { 80 hollow | | | | 171.5 | | 5.4 | 6.7 | " |
| 104 | } stay bolts. | | | | 87 | | 6.94 | 8.6 | " |
| 105 | No. 150......... | 15.3 | 787 | 51.4 | 100 | | 5.71 | 7.1 | " |
| 106 | No. 159......... | 13.1 | 801 | 61.3 | 95 | | 5.75 | 7.2 | " |
| 107 | No. 171......... | 15.3 | 839 | 54.0 | 71 | | 6.28 | 7.9 | " |
| | L. & N. | | | | | | | | |
| 108 | No. 255......... | 12.5 | 1141 | 91.2 | 103 | Pittsburgh best.. | 8.77 | | " |
| 109 | | | | | 111 | "   poor. | 6.98 | | " |
| 110 | | | | | 115 | Alabama......... | 7.68 | | " |
| 111 | | | | | 130 | Tennessee....... | 7.52 | | " |
| 112 | | | | | 146 | Cent. Kentucky.. | 6.47 | | " |
| 113 | | | | | 127 | "    " | 6.58 | | " |
| 114 | | | | | 113 | "    " .. | 7.60 | | " |
| 115 | | | | | 111 | "    " .. | 7.80 | | " |
| 116 | | | | | 145 | "    " .. | 6.11 | | " |
| 117 | | | | | 113 | W. Virginia...... | 7.79 | | " |
| 118 | | | | | 120 | E. Kentucky..... | 7.24 | | " |
| 119 | | | | | 133 | " ..... | 6.80 | | " |
| | Torpedo boat. | 18.9 | 618 | 32.7 | 48.9 | | 8.88 | | |
| | Fan draft. | | | | 62.2 | | 8.26 | | |
| | | | | | 78.9 | | 7.94 | | |
| | | | | | 96.0 | | 7.48 | | |

## CHAPTER III.

### EXTERNALLY FIRED STATIONARY BOILERS.

Boilers may be defined as the closed vessels in which steam is generated from water by the action of heat applied from the outside,—water being introduced at one aperture and steam removed at another. The heat is usually the heat carried in the products of combustion, or hot gas from a furnace, which is passed along the surface of the boiler and by giving its heat to the material of the boiler first heats that material, that again heating the water in the boiler.

The shapes which have been given to boilers are almost endless and it is not our intention to attempt any complete collection of the types which have been used and abandoned, or are still being experimented with; but, on the contrary, our aim has been to exclude all but what may be called the standards, such as have been used for years with satisfaction.

Modern boilers are usually either cylindrical, or a combination of cylindrical with rectangular forms. The material is either wrought iron or steel. Copper is still used in the fire-boxes of English locomotives, and brass tubes are also in use. The strength and capacity we shall discuss later, and at present content ourselves with a classification and brief description of the standard types.

In classifying boilers we may base our system upon their use, whether for stationary, locomotive, or marine purposes, and this gives a very convenient division. Stationary boilers may be divided into three primary groups.

*First.*—Cylinder forms with the fire external and the products of combustion also external to the cylinders.

*Second.*—Externally fired boilers with fire tubes, by which the products of combustion are passed through the water or steam.

*Third.*—Internally fired boilers in which the furnace is inclosed by a water chamber on all sides but one or two, and the products of combustion pass through tubes surrounded by water.

Locomotive boilers, including with them portable boilers, are usually internally fired, having a rectangular furnace with fire tubes passing through a horizontal shell; sometimes portable boilers are made with vertical cylindrical shells and fire-boxes, with vertical fire, and sometimes also with vertical water, tubes. There is less variety in locomotive boilers proper than in any other class; with marine boilers, on the other hand, the modifications have been almost endless.

The types of marine boilers are now much better defined and may be considered as established, and we have two broad divisions, being the practice in and around New York and the practice on the Clyde.

Marine boilers were for many years after their first introduction limited to low pressures on account of their being fed with salt water from the hot well of an injection condenser, and as long as the pressure was not over twenty pounds per square inch, they could be kept clean by blowing out and washing; with a higher temperature, the water deposited salt too rapidly and made scale too fast. On the Western Rivers, where fresh but very bad water was to be had, the use of the condensing engine was soon abandoned, and by the use of steam at pressures exceeding 150 pounds per square inch, a very simple and cheap engine and boiler were developed which is at least as economical in the use of fuel as the low pressure condensing engines of that day. The practice around New York has remained almost stationary, while that of the Clyde with the introduction of the surface condenser began at once raising the pressure and introducing compound engines, until in 1882 several steamers were set at work with 125 pounds pressure per square inch, and the magnificent steamers of the Atlantic lines are working with ninety pounds. The introduction of higher pressures required: first, stronger boilers, and then more accessible boilers for the removal of harder scale and for more perfect inspection; and the Clyde practice gradually grew definite.

The North River Boiler, as the New York type is called, has one or more rectangular internal furnaces open on the bottom and with a cast-iron front; the products of combustion are carried through large tubes to a connecting chamber and then, ascending, pass back in smaller tubes to the front of the boiler and through a shell of steam in an annular chamber around the base of the stack. The external form is a cylinder with a rectangular block inserted at the bottom of one end. The use of flat-stayed surfaces around the fire separates this type from the Clyde type, which uses cylindrical furnaces as well as shell. The use of fire tubes of large diameter has always been confined to the English practice, and in this country we never possessed, until very recently, the facilities for manufacture which their usage demands. The Clyde boilers may be classed by the number of furnaces, from one to four, and by their being fired from one or both ends. Locomotive boilers are used for high pressures on the torpedo boats, and all the forms of upright boilers used with steam fire engines have been used with small boats, as also the water tube types, especially that of Herreschoff, in connection with surface condensers.

Externally fired boilers are usually combinations of cylindrical forms set with axes either horizontally, vertically, or inclined. Of the types with horizontal axis, the simplest is the plain cylinder boiler set in brickwork, either singly or in groups of two or more, called batteries. The only objections are the room taken up, the amount of land occupied, and the quantity of brickwork, with its liability to leak air into the furnace and consequently the reduction of temperature therefrom. The boiler is usually hung from overhead supports and the brickwork held in place by tie rods passing over the boiler connecting vertical bars called "buck staves," these act as anchor plates against the movement of the links outward under the action of heat.

The advantages of the single cylinder boiler are simplicity, ease of access for inspection and repairs, strength, durability, and low cost of construction, which render this class of boiler well adapted to hard, continuous, and high pressure work. They are commonly used for blast furnaces where land is cheap and it is desirable to keep them in steam for a long time. They are usually fired with the waste gas from the blast furnace, but must, of course, have independent grates to use when the furnace is cold at starting or "blowing in;"—the gas when turned over the existing fire and supplied with enough air burns freely.

We select as an example the boilers of the Meier Iron Company, at Bessemer, Illinois, opposite South St. Louis, constructed upon the following specifications. There are five batteries of two boilers each.

SPECIFICATION FOR ONE BATTERY OF STEAM BOILERS FOR THE MEIER IRON COMPANY.

Each Battery is to consist of two plain cylindrical boilers with rounded ends, one steam drum and two mud drums, each boiler to be 60 feet long and 42 inches in diameter. The boilers, drums, legs, and other parts to be built and suspended according to the drawings. The shells of boilers to be of $\frac{1}{4}$-inch iron and the heads of $\frac{1}{2}$-inch iron. There is to be one elliptical manhole in each boiler of sixteen (16) and ten (10) inches diameters bound with a wrought-iron gasket of 5-inch by $\frac{3}{4}$-inch and closed with a cast-iron plate and bridge held down on a lead gasket by wrought-iron bolts. The centre of these manholes to be exactly in the upper middle lines of the boilers and in the middle of the fourteenth sheet from the grate end of the boilers.

*No Horizontal* seams to be below the fire tiles.

*Steam Drum* to be 24-inches diameter, shell of $\frac{1}{4}$-inch, heads of $\frac{3}{8}$-inch boiler iron, legs of $\frac{1}{4}$-inch iron. There is to be a manhole in one end, elliiptical, 10-inches and 14$\frac{1}{2}$-inches diameters fitted up and closed like those in the boilers. There are to be *two mud drums*, 24 inches diameter. Shells of $\frac{1}{4}$-inch boiler plate. Heads of 1$\frac{1}{2}$-inch cast-iron. Each to have a manhole at one end of same size, and fittings as on the steam drum.

Below the manhole in the mud drums there is to be a blow-off, or mud valve, 4 inches in diameter. The other end of each mud drum is to have a 4-inch check valve, with flange for attaching to the feed pipe. Each battery is to be suspended in four places by bolts and supports consisting of channel bars carried on rails as shown in the drawings, and the boilermaker shall erect these supports and suspend the boilers.

The plate iron of the boilers and drums throughout is to be all C. H. No. 1., with maker's brand designating the quality stamped upon each sheet, and each sheet must be guaranteed to have a tensile strength of fifty thousand (50,000) pounds per square inch. The Meier Iron Company to have the right to require clippings of any sheet made, for the purpose of testing the same.

The boilers when fitted to be tested to 100 pounds per square inch of hydrostatic pressure.

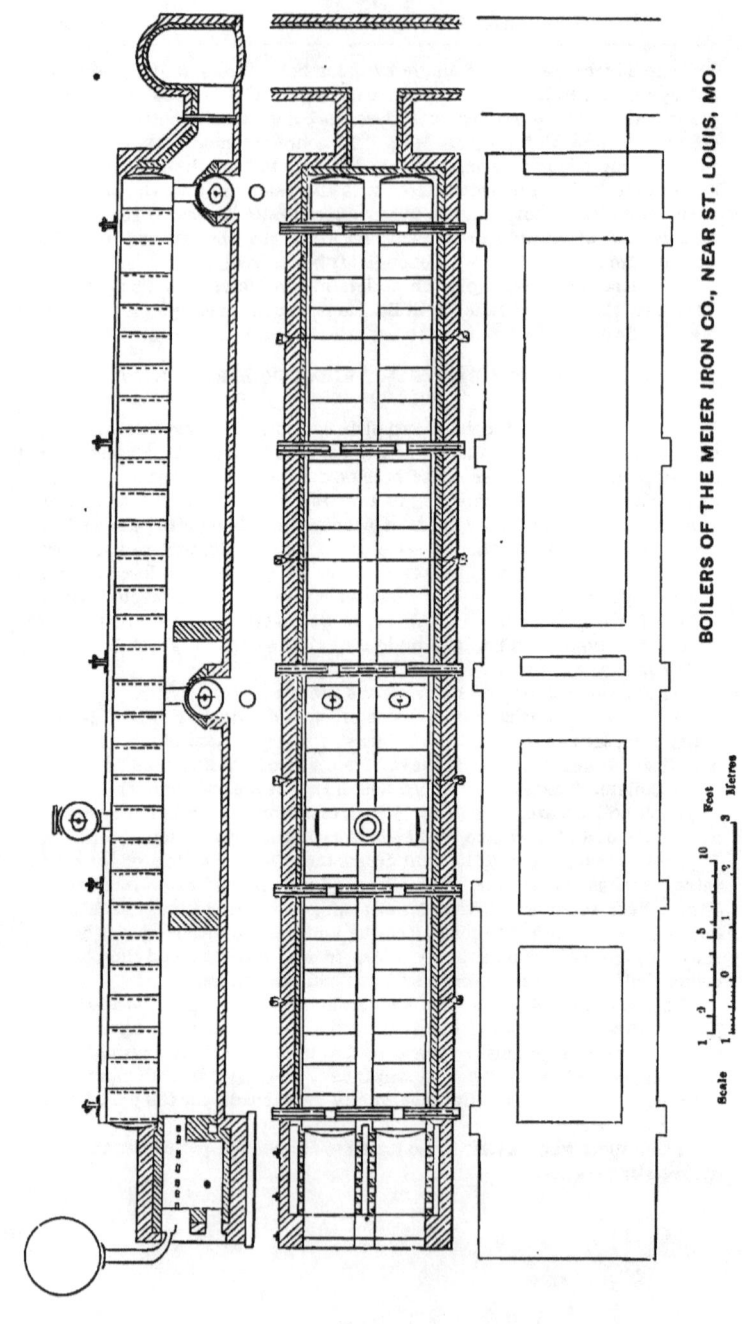

BOILERS OF THE MEIER IRON CO., NEAR ST. LOUIS, MO.

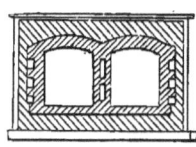 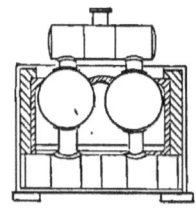

TRANSVERSE SECTIONAL VIEWS.

SPECIFICATIONS FOR FITTINGS FOR BOILERS FOR MEIER IRON COMPANY.

*I. Safety Valves.*—There is to be one on each steam drum of 4 inches clean diameter and to have a lever and weight to counter-balance seventy-five pounds per square inch pressure.

*II. Steam Valves.*—One 8-inch globe valve for each battery of boilers attached by means of flanges and bolts to the cast-iron pipes between steam drums and 20-inch steam pipe.

*III. Blow-off on Mud Valves.*—One 4-inch valve for each mud drum attached to a neck and flange cast on the head of the drums.

*IV. Glass Tube Water Gauges.*—One to each boiler so arranged that the glass tube can be removed or replaced while the steam is on the boiler. Tubes 12-inches by ⅜-inch best Scotch glass, fittings to be screwed into one head of the boiler.

*V. Mississippi Gauge Cocks.*—Three to each boiler in the head, to be hereafter designated by the Meier Iron Company. Shank one inch diameter.

*VI. Steam Whistle.*—One 5-inch whistle placed on the main steam pipe at a point to be hereafter designated.

*VII. Steam Gauges.*—Two John Kupferle & Company steam gauges, 7-inch face for pressures up to 150 pounds per square inch, brass casings with syphon pipes carefully tested. One of these to be placed on the 20-inch steam pipe and one in the engine-house at points to be hereafter designated.

*VIII. Check Valves.*—One 4-inch check valve with flange to attach feed pipe for each mud drum. All the above pieces to be of the best material and workmanship, all joints perfectly steam and water tight, and subject to inspection by the experts of the Meier Iron Company. The contractors to fit up, fasten, and erect all parts ready for service.

*Accepted by*..................................

The shells are made with taper rings thus making the outside lap always at the rear end of the sheet. The boilers are set in a straight flue

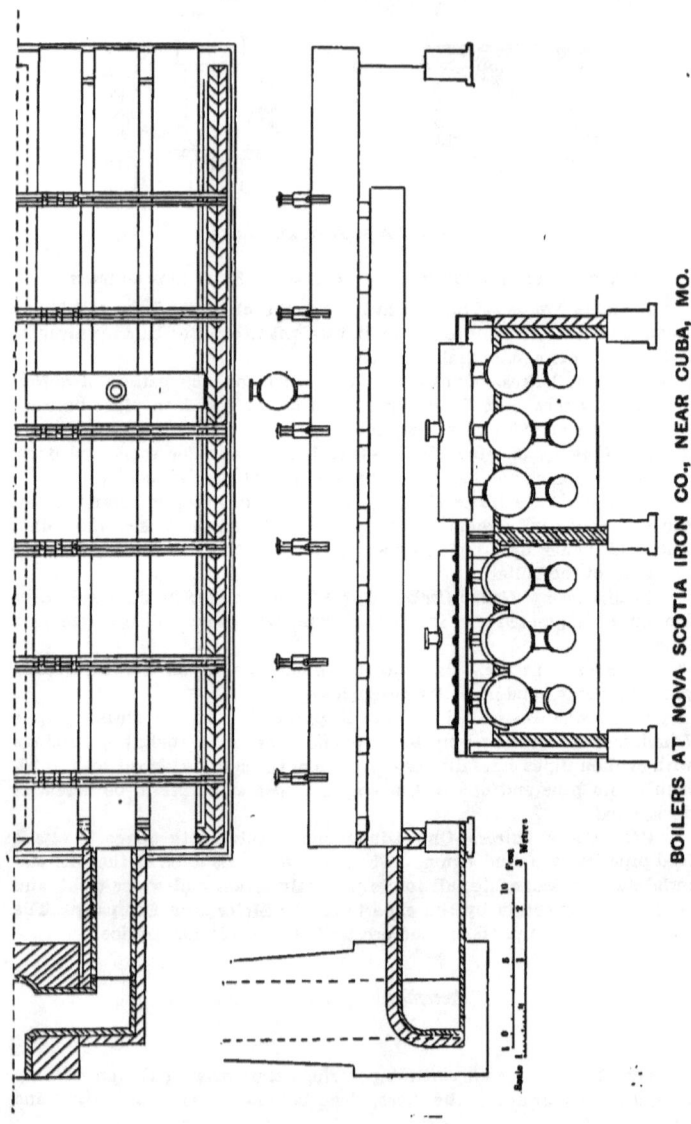

with a fire brick lining. There is a combustion chamber in front of each boiler where the gas from a 60-inch overhead pipe receives air through side openings in the chamber, and is ignited. Two fire-brick bridges were added afterwards. The flue is floored with red brick and the mud drums are protected from the flame by fire-brick arches open through the side walls. The rear end of the flue connects with an underground flue to a stack 204 feet high and 10 feet 6 inches diameter of opening at the top.

These boilers were designed by Mr. E. D. Meier, a member of the Board of Direction.

In point of simplicity and durability a combination of two shells comes next, and our example is taken from the boilers at the works of the "Nova Scotia" Iron Company, near Cuba, Missouri.

There are two batteries of three boilers each, side by side, fired by gas from the furnace. Each boiler is carried on the front, and by six sets of slings, and consists of two shells one over the other.

The upper shell is $40\frac{1}{2}$ inches in diameter and 55 feet 8 inches long, and the lower shell is 30 inches diameter and 45 feet 8 inches long. The shells are $16\frac{1}{2}$ inches apart and are united by nine 13-inch legs with cap flanges 5 feet 4 inches from centre to centre. The back ends of upper and lower shells are in the same vertical plane, and the lower one being 10 feet shorter than the upper, leaves a combustion chamber at the front end.

The slings are made by a 4-inch tee-iron riveted on the upper side of the shell, forming nearly a semi-circle, and three bolts 1-inch diameter with nut at the lower end and thread and nut at the upper. These are passed between two 10-inch I beams which rest on cast-iron "buck staves" built in the sidewalls.

The metal is all $\frac{1}{4}$-inch in thickness.

When there are two or three smaller lower cylinders in place of the one, the boiler is known in England as the "French" or "Elephant" boiler, and is called in France "Chaudiére a Bouilleurs," or boiler with heaters, the latter term being applied to the smaller lower cylinders. The objections to these combinations is the weakening of the larger shell by the openings for the legs, and the straining action produced by unequal expansion; but as the metal used is very thin, the boilers are very durable. The only example we know of in the United States of the French boiler is a small one with two "heaters" which had been in use for nineteen years at the shops of the St. Louis, Alton & Terre Haute Railroad Company, in East St. Louis. The service was very moderate, but in that period the only care taken was to wash out once a week. The metal was $\frac{3}{16}$ inch in thickness and the shells were not over 30 inches and 10 inches diameter, if our memory is correct.

The "French" boiler tried at Mulhouse at the Works of the Société Alsaciennes des Constructions Mechanique, in 1875, by a Committee of the Société Industrielle de Mulhouse, was of the following dimensions: Shell, 29 feet $6\frac{3}{4}$ inches long, by 3 feet $8\frac{9}{10}$ inches diameter by $\frac{1}{2}$-inch thick, in eight rings, with rounded heads, $0\frac{55}{100}$ inch thick. Heaters, were three in number, each 32 feet $9\frac{3}{4}$ inches long by $19\frac{7}{10}$ inches diameter, by $0\frac{39}{100}$ inch

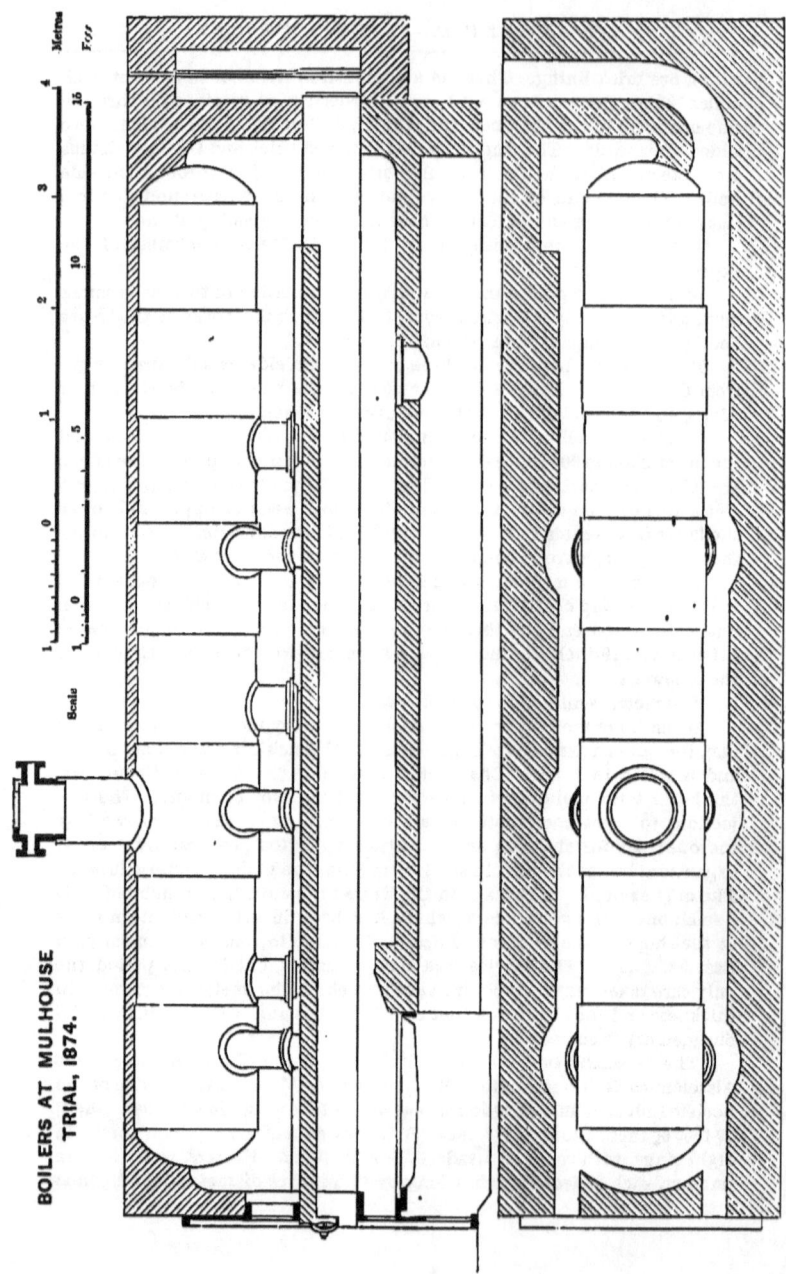

BOILERS AT MULHOUSE TRIAL, 1874.

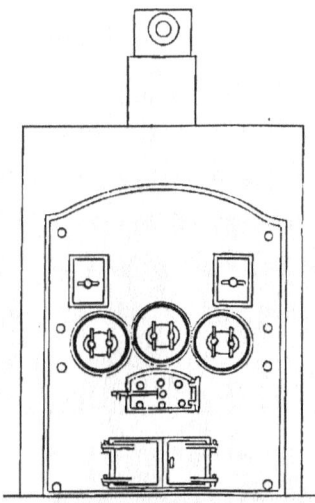
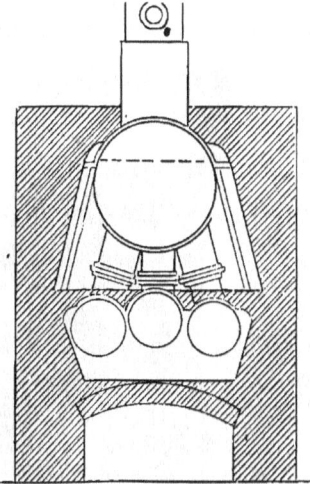

END ELEVATION.   TRANSVERSE SECTION.

thick, each connected by three legs about 14 inches diameter. Those for the inner heater are about 12 inches long and those for the outer heaters about 21 inches long on the centre line.

The grate is 4 feet $9\frac{1}{10}$ inches wide by 4 feet $9\frac{3}{10}$ inches long, of which $6\frac{7}{10}$ inches is over the bearing bars. The effective length being 4 feet $2\frac{6}{10}$ inches, giving an area of $20\frac{95}{100}$ square feet. The heating surface is as follows:

|  | Square feet. |
|---|---|
| Main shell | 199.5 |
| Heaters | 385.7 |
| Legs | 22.4 |
| Total | 607.6 |

The products of combustion pass first from the grate over a low bridge along the "heaters," then returns to the front on one side of the shell, and returns to the rear along the other side of the shell.

The only thing we note is the great thickness of the plates used, which, with bad water foaming, causes much scale to invite burning. The United States law thus limits the thickness of plates exposed to the direct action of the fire for steamboat boilers: externally fired to $0\frac{25}{100}$ inch, and as such boilers are frequently found in excellent order after 20 years of service, with pressures exceeding 100 pounds per square inch on shells 42 inches in diameter, single rivetted, there seems little use in making them ½-inch thick to carry 70 pounds or so per square inch. Our illustration shows this boiler and setting.

From the "French" to the water tube boilers is a natural transition, the water cylinders growing smaller and more numerous till the well known Howard, Belleville, Root, Kelly, Firminich, Perkins, Babcock & Wilcox, and Heine boilers were introduced. In most of these the greater com-

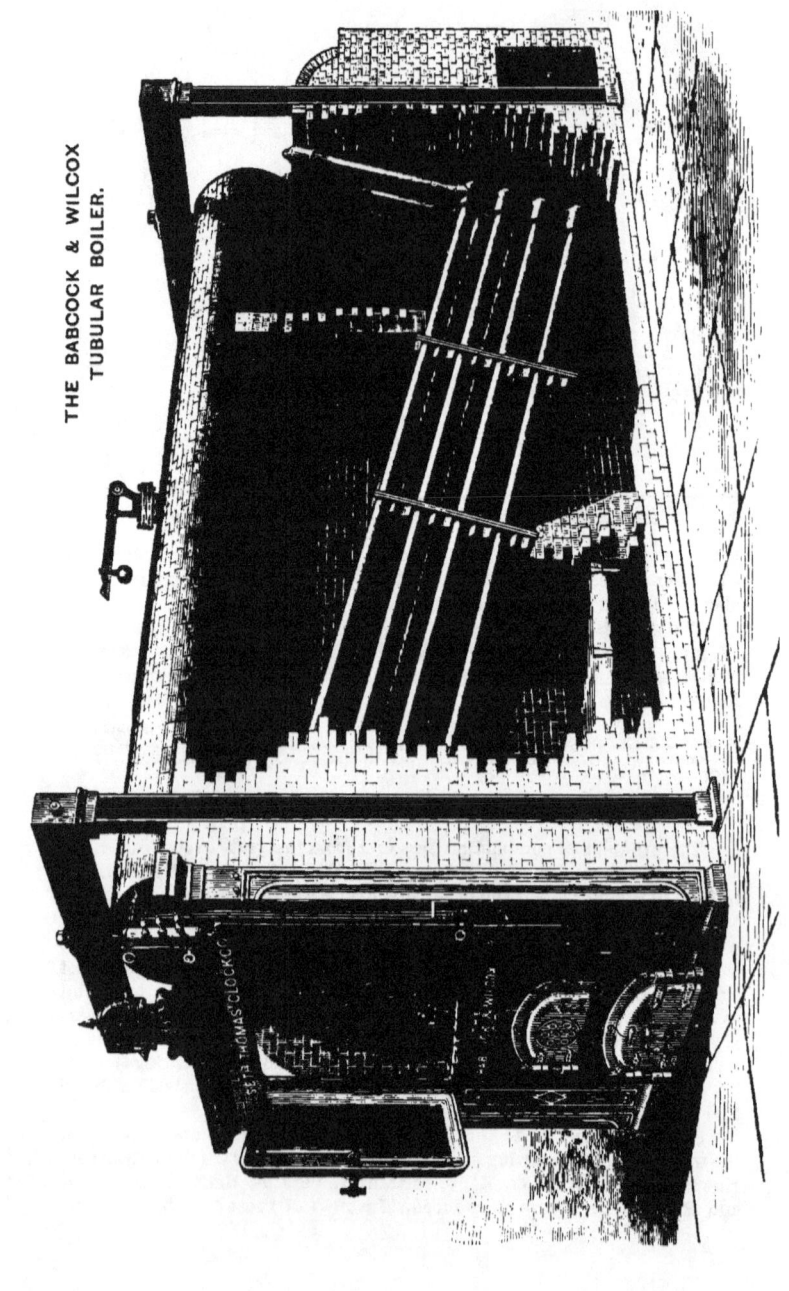

THE BABCOCK & WILCOX TUBULAR BOILER.

plexity of arrangements for cleaning seem to balance any advantages they may have in circulation, and the leakage at the innumerable joints goes far to neutralize any gain made by the higher steam pressure which can undoubtedly be carried with safety.

Of the many varieties of the water tube type we select as example one from the circular of the Babcock & Wilcox Company.

The Babcock & Wilcox boilers consist of inclined sets of tubes connected at each end by steel castings into which the tubes are expanded: the front and upper ends are connected into a longitudinal drum. The rear ends are connected by inclined pipes to the same drum, while at the bottom there is a mud drum of cast-iron.

There are hand holes opposite each end of each tube and man heads on the steam and mud drums. The hand hole plates are milled to metal contact with the connections and the whole section is hung from overhead by bolts carried on cross-beams resting on the walls. The products of combustion are carried three times across the tubes by means of deflectors and these from all the different batteries are carried through the pipes of an "economiser," or feed-heater, on their way to the stack.

In Heine's boiler the connections at the front and back ends of the tubes are flat-stayed wrought-iron plates, the stay bolts are made hollow and plugged; by taking out the plugs the outside of the water tubes may be cleaned. There are hand holes and plates opposite each end of each tube. The inclination of the tubes is not so great as in the other boiler. There are many other varieties of the water tube boiler, from the single coil of Jacob Perkins, used with a "separator" and circulator by John Elder, and afterwards revived by Herreschoff, to the network or cage of tubes used by Loftus Perkins.

A very long time ago the device of an internal flue was introduced to increase the heating surface without increase of ground room, and an advantage was at once gained thereby,—that with less water and more surface, steam could be raised more quickly: at first a single tube was used, then two or more of smaller diameter, till finally the multi-tubular boiler provided great heating surface within moderate space.

Of the externally fired boilers with return fire tubes, or flues, the use of a single return flue is not common in the United States, and we have no knowledge of their use except in some cases where the waste heat of puddling furnaces has been utilized; also a few which went out of service twenty years ago.

Horizontal externally fired return flue boilers, with from two to twelve flues, are the almost universal boiler of the Mississippi Valley, a natural consequence of their use upon the rivers for steamboats, and we shall consider them here as stationary boilers.

The chief restrictions upon the construction of boilers for use in steamboats upon the Mississippi River and tributaries are as follows:

1. Boilers must be tested at least once a year by hydrostatic pressure and the test applied must exceed the working pressure allowed in the ratio of three to two.

2. Fire line must be kept at least 2 inches below low water line.

3. Water level must be kept not less than 4 inches over flue.

4. Feed-water must be so delivered as not to injure boiler when entering it.

5. Fusible plugs must be placed in such position as to melt when water gets too low.

6. Boilers 42 inches in diameter and $\frac{1}{4}$-inch shell may be allowed a working pressure of 150 pounds per square inch and the pressure allowed for other diameters and thicknesses may vary inversely as the diameter and directly with the thickness. If double rivetted longitudinal seams are used the pressure allowed is 20 per cent. more.

7. Each plate must be stamped with the number of pounds tensile strain it will bear.

8. The working pressure allowed must not exceed one-sixth of the tensile strain of the sheets, unless the longitudinal seams are double rivetted, in which case 20 per cent. additional may be allowed.

9. The plates of externally fired boilers exposed to the action of heat must not be over $0\frac{25}{100}$ inch thick.

10. The flues or tubes of externally fired boilers must have not less than 3 inches clear space between and around them.

For the two-flue type we select the boilers of the river steamer "Montana," as described by Mr. Wm. H. Bryan, Mechanical Engineer, of St. Louis, Mo.:

The boilers are four in number, set in one battery. They are each 42 inches diameter and 26 feet long, with two flues, each 15 inches diameter. They are of C. H. No. 1 iron, and were built by D. W. C. Carroll, of Pittsburgh, Pa. The shell is of $0\frac{25}{100}$ wrought-iron in sheets 24 inches wide and 1 inch lap on circumferential seams, single-rivetted 1$\frac{3}{4}$-inch centres; longitudinal seams, double rivetted 1$\frac{3}{8}$-inch centres, rows 1$\frac{3}{4}$ inches apart. Heads of $\frac{1}{2}$-inch iron, C. H. No. 1. Flange 3 inches inside shells, rivets 1$\frac{3}{4}$ centres. Steam drum is 20 inches diameter by 15 feet long, and connects with each shell by a 14-inch leg. Centre of drum 37 inches above centre of shells. Back head flanged in for flues. Front end flanged outward for same. Shell sheets alternately lap in and out, making laps all toward after end. Flues all lap same as shells, thus the gas never strikes fair on a caulked edge. Flues of $\frac{1}{4}$-inch iron 15 inches diameter and centre of flues 18 inches apart and 3$\frac{1}{2}$ inches below centre of boiler. One manhole 9$\frac{1}{2}$ inches by 15 inches above flues at after end of each boiler: its lower edge is 13$\frac{1}{2}$ inches from top of boiler and $\frac{1}{2}$-inch below low water line. Manhole strengthened by $\frac{3}{8}$-inch ring inside. Manhead plate of cast-iron with two arch bars. Hand-hole 4 inches by 6 inches in front head, lower edge 2 inches from bottom. Heads stayed to shell. There are two mud drums, one under second sheet from after end and one under fifth sheet from front end. Steam drum over fifth sheet from after end. Feed water introduced at after drum. Mud drums are 16$\frac{1}{2}$ inches inside diameter and 15 feet long, with centre 41 inches below centre line of boiler, united to each shell by 8-

inch legs. Steam is taken from steam drum by a 6-inch copper pipe. Safety valve on each boiler with area 11 square inches. Lever is 4 feet 3 inches long, notched at intervals. Weight is 9½ inches by 9½ inches by 8 inches and weighs 200 pounds. Blows off at 140 pounds per square inch and sets weight at 31 inches from fulcrum. One pressure gauge is placed at the front of the boilers and one near the engines. There are 10 gauge cocks on the four boilers and each boiler has a float gauge with outside dial. These are at the after end of the boiler close to the feed pump and are under the eye of the engineers on watch.

The front end of the boiler is carried by the cast-iron front. A 3-inch by ½-inch ring is rivetted on the shell and rests on the front. The other end of the boilers is carried on the back mud drum, which rests on cast-iron supporting blocks.

The furnace is 14 inches in height from the grate bars to the shell and 37 inches high between boilers; it is 17 feet wide under all four shells, 6½ feet long to top of bridge wall and is built with fire brick throughout. The grate is 17 feet wide by 4 feet 2 inches long and 70.8 square feet in area. The bars are double, of cast-iron, with 1-inch air spaces between the pair of bars, and with ½-inch lugs, giving the same space to the next bar. The top of grate is 2 inches below lower lining of fire-door. The doors between shells are 18 inches wide by 14 inches high, with half doors 12 inches wide and 13 inches high on outer side of outer shells. The doors are of cast-iron with ½-inch holes and the front is of cast-iron, in pieces, bolted together, and lined with fire-brick. The ash pit is 18 inches below bars and the same area as the grate. The drip from the long exhaust pipe is run in here to put out the fire and cool the ashes which fall through the bars. The ash-pit doors are of sheet-iron, three large and two small ones. The fire bridge wall is 11 inches in height with a slope up from the grate thereto, a run horizontally of 2 feet, carried with the rear end of the grate by a special frame. The flame chamber slopes from the bridge, where it is 3 inches below the shell, to the rear end, where it is 6 inches below. There is a lining of 4-inch red brick set in and covered by clay, and the side walls are made in the same way, carried in a sheet-iron casing, and supported from the deck by iron rods.

The stacks are two in number, each 3 feet in diameter, and 55$\frac{3}{6}$ feet above the grate, and are of No. 12 iron. The "breeching" or smoke connection is of the same thickness, and is provided with doors opposite the flues in each boiler.

The proportions and dimensions are summed up:

Grate area.................................................................. 70.8 square feet.
Total heating surface......................................................1431.2 " "
Ratio, about...............................................................20 to 1
Calorimeter............................................................... 9.82 " "
Grate area to Calorimeter................................................. 7.2 " "
Area stacks............................................................... 14.14 " "
Grate area to stack area.................................................. 4.6 " "
Steam room................................................................ 562 cubic feet.
Water room................................................................ 294 " "
" " ...................................................................... 2208 gallons (U. S.)

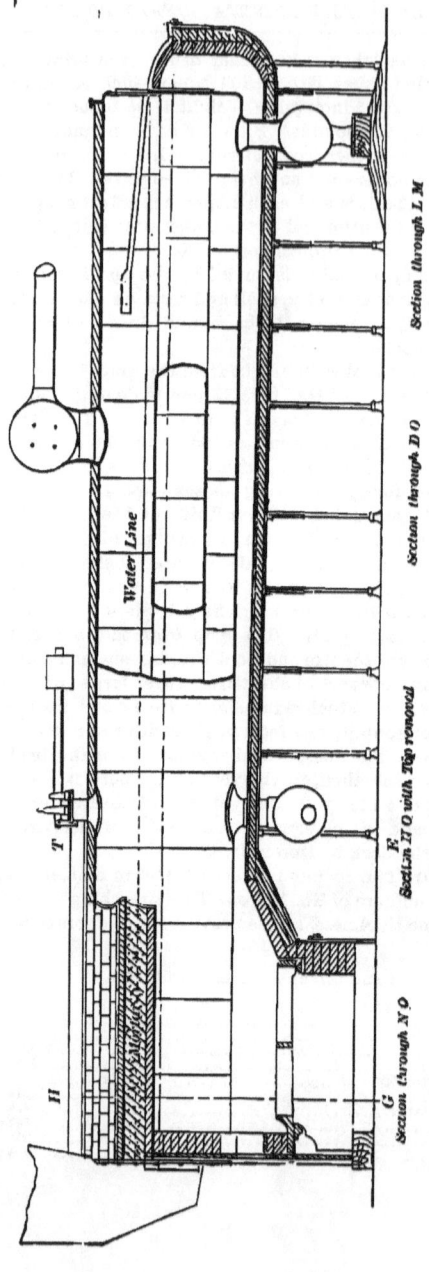

WESTERN RIVER STEAMBOATS—BOILERS OF THE "MONTANA."

# EXTERNALLY FIRED STATIONARY BOILERS.

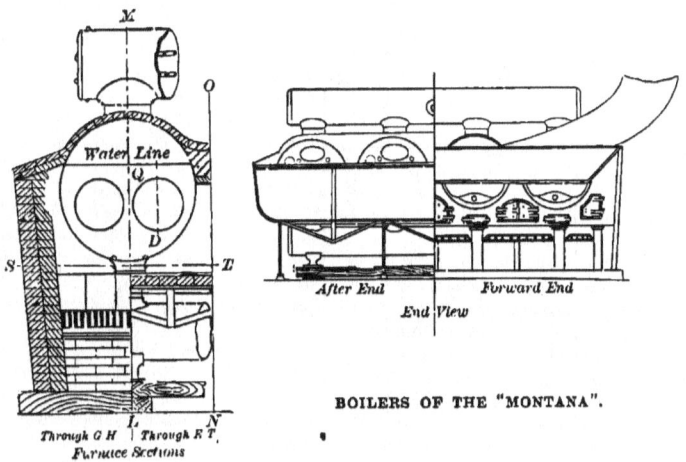

BOILERS OF THE "MONTANA".

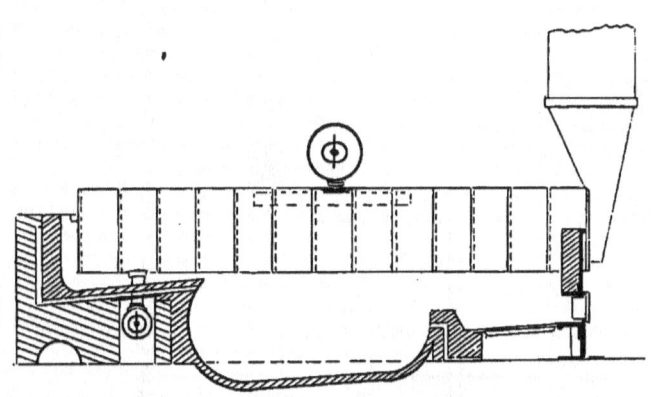

BOILERS AT THE LA CLEDE ROLLING MILL,
ST. LOUIS, MO.—Elevation.

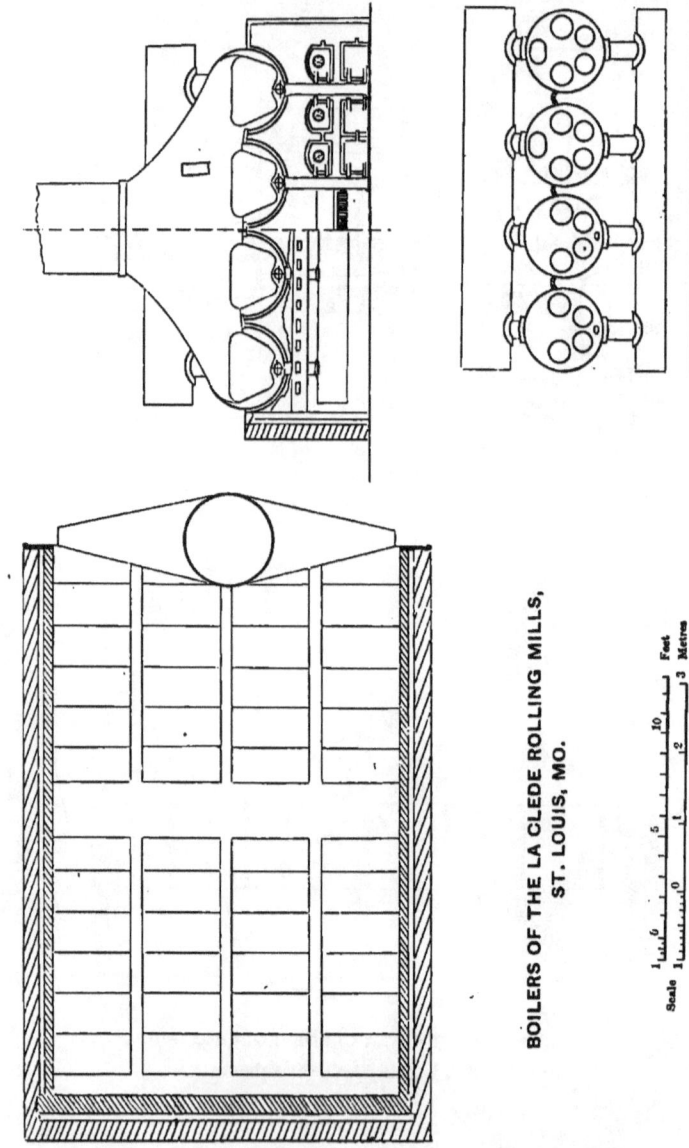

BOILERS OF THE LA CLEDE ROLLING MILLS,
ST. LOUIS, MO.

| | | |
|---|---|---|
| Weight of boilers, etc. | 29263 | pounds. |
| Weight of water. | 18350 | " |
| Weight of stacks | 2357 | " |
| Weight of grate | 5700 | " |
| Total | 55671 | " |

Next in point of simplicity come boilers with four flues, and we take as an example four boilers built in 1881 for the Laclede Rolling Mill, St Louis, which were built under the following specification:

### SPECIFICATION FOR A BATTERY OF FOUR BOILERS FOR LACLEDE ROLLING MILL.

ST. LOUIS, MO., Aug. 1, 1881.

*Shells.*—The shell to be of the best hammered charcoal iron, 48 inches diameter, 26 feet long, and $\frac{1}{4}$-inch thick, and to have a tensile strength of 55,000 pounds per square inch. All longitudinal seams to be above the fire line and double rivetted. Seams to be staggered to prevent a continuous row of rivets.

*Heads.*—Heads to be of the best hammered charcoal flange iron, and to be $\frac{5}{8}$-inch thick. The back head of each boiler to contain an 11 inch by 15-inch man-hole, and the front head a 4-inch by 6-inch hand-hole.

*Flues.*—Each boiler to have two 12-inch flues and 11-inch flues, as shown on drawing. Flues to be of $\frac{1}{4}$-inch iron and of same quality as shell.

*Steam Drum.*--The steam drum to be 30 inches in diameter and the length equal to the width of the battery of boilers. The legs of steam drum to be 10 inches diameter and 8 inches long. Each end of steam drum will contain an 11-inch by 15-inch man-hole. The iron used for steam drum and legs to be $\frac{1}{4}$-inch thick and of same quality as that used in shells. Legs to be attached by cap flanges.

*Mud Drum.*—The mud drum to be at back end of boilers, as shown on drawing, and to reach the entire width of the bottom. Diameter of drum to be 18 inches, and of same quality of iron as shell. Legs to be 8 inches diameter and attached by cap flanges. Thickness of iron for mud drums and legs $\frac{5}{16}$-inch. Length of legs, 20 inches. Each end of mud drum will contain a 10-inch by 14-inch man-hole and one end to be furnished with a 3-inch brass blow-off cock. Heads of mud drum $\frac{3}{8}$-inch thick.

*Supports for Boilers.*—The front ends of boilers will rest on fire front and the back ends will rest on cast-iron stands.

*Safety Valves.*—There will be two $4\frac{1}{2}$-inch safety valves, one near each end of steam drum, and fitted with proper weights and levers for a pressure of 100 pounds steam. The chests to be of cast-iron, the valves and seats of composition, as per drawing.

*Dry Pipes.*—A dry pipe $7\frac{1}{2}$-inches diameter, and 8 ft. long to be placed in each boiler near the top. The top of dry pipe will be drilled with holes $\frac{3}{8}$-inch diameter, equally distant apart; aggregate area of holes to be twice the area of the pipe.

*Stop Valve.*—One 12-inch iron globe valve with outside screw steam valve and seat to be of brass, to be secured to steam drum by flanges.

*Gauge Cocks.*—Three ⅜-inch Mississippi gauge cocks to be placed in back end of wing boiler.

*Glass Water Gauges.*—One glass tube water gauge, to be placed in the front end of each wing boiler.

*Breeching.*—The breeching to be of No. 12 iron, with doors opening upwards, provided with hinges and latches.

*Boiler Front.*—The fire front will have tight-fitting fire and ash pit doors, with suitable hinges and latches. Registers and perforated lining onfire doors, as per drawing. Top of grate bars 30 inches below boilers.

*Stack.*—The stack will be 60 feet high above breeching, and 4 ft. 6 in. diameter, with damper above breeching.

*Man-holes.*—The man-holes to be 11 × 15 inches, each man-hole to have around it on the inside an elliptic ring of iron 2½ × ¾ inch. Rivets to be countersunk flush on both sides. Parts where gasket joint comes to be faced. Man-hole plates to be of cast-iron, and secured by wrought-ron arches and bolts. A 2 × ½ inch wrought iron ring on outside of hand-holes.

*Steam Gauge.*—One Ashcroft's steam-gauge, 8-inch face.

*Bearing Bars.*—Bearing bars to be furnished as per drawing. Grate bars to be furnished by company.

No paint or putty is to be put on any part of boilers until the same have been delivered and tested at the works. Only first-class workmanship will be accepted.

The setting of these boilers is in red brick, lined with fire brick. There are two bridges, and air is admitted through the rear bridge to the combustion chamber, which is rather peculiar in form.

Boilers with five flues are very much used in the Valley of the Mississippi River, but there is little difference between them and those last described.

What is known as a compromise between flue and tubular boilers, is one with 6-inch lap-welded tubes. Of this class, we give as an example, a battery of three boilers at the works of the St. Louis Lead and Oil Company, St. Louis. These boilers are each 42 inches diameter, and 22 feet long. Shells of ¼-inch thick C. H. No. 1 iron, single riveted. Each boiler has eight 6-inch lap-welded tubes, with steam and mud drums, as shown in the drawing, which requires no further explanation. The heads are flanged and the flues riveted thereto.

As an example of the 4-inch tubular, we describe three boilers designed by us for Washington University, and erected in 1879. They are set independently and are used mainly for heating.

Each boiler is 60-inches diameter, and 16 feet long, with 36 lap-welded tubes, 4-inches external diameter arranged in two groups, with a central gangway 12 inches in the clear in the centre of the boiler. This space allows circulation and complete inspection, with convenience in cleaning. The shell is ¼-inch steel, the heads ½-inch steel with tensile strength of 70,000 pounds per square inch. The shell is supported by the front be-

# EXTERNALLY FIRED STATIONARY BOILERS. 65

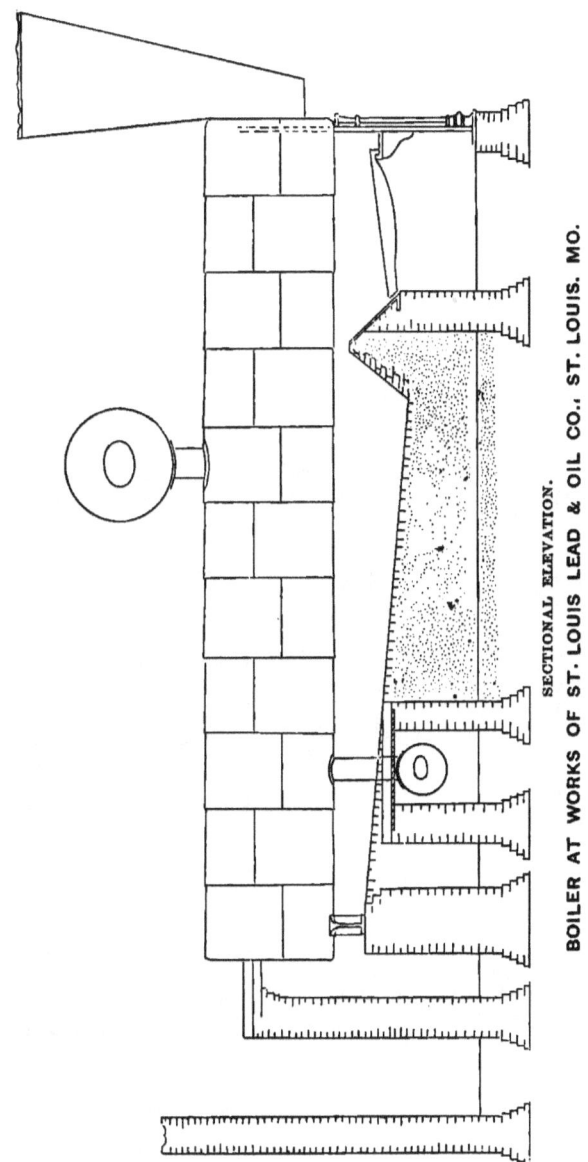

BOILER AT WORKS OF ST. LOUIS LEAD & OIL CO., ST. LOUIS, MO.
SECTIONAL ELEVATION.

66 STEAM MAKING; OR, BOILER PRACTICE.

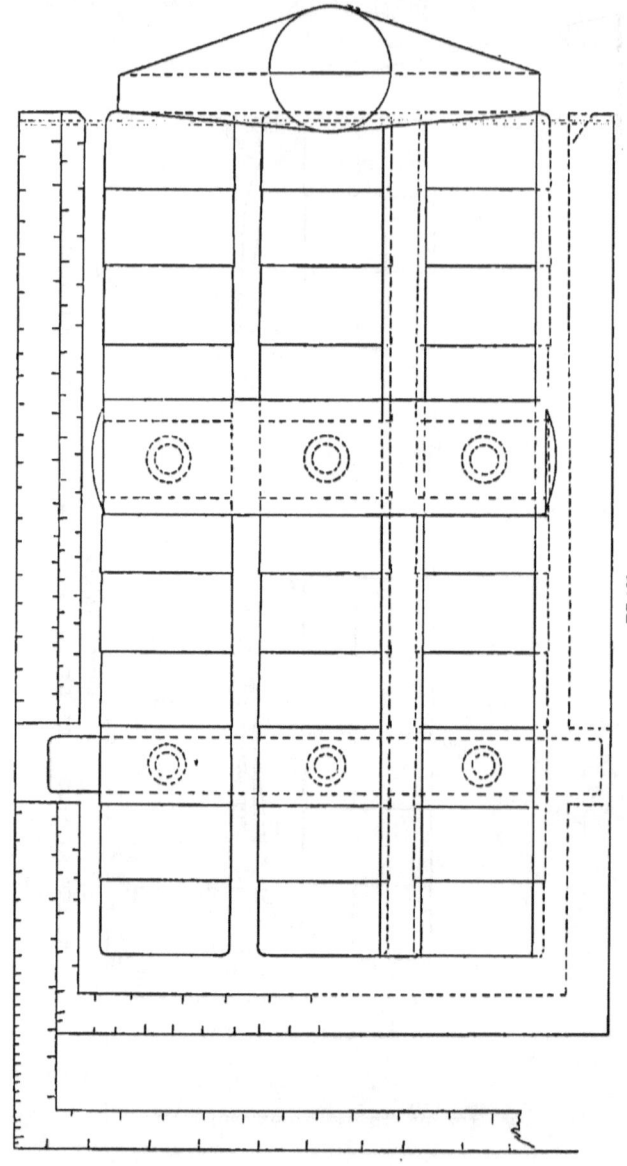

PLAN.

BOILERS AT WORKS OF ST. LOUIS LEAD & OIL CO., ST. LOUIS, MO.

# EXTERNALLY FIRED STATIONARY BOILERS. 67

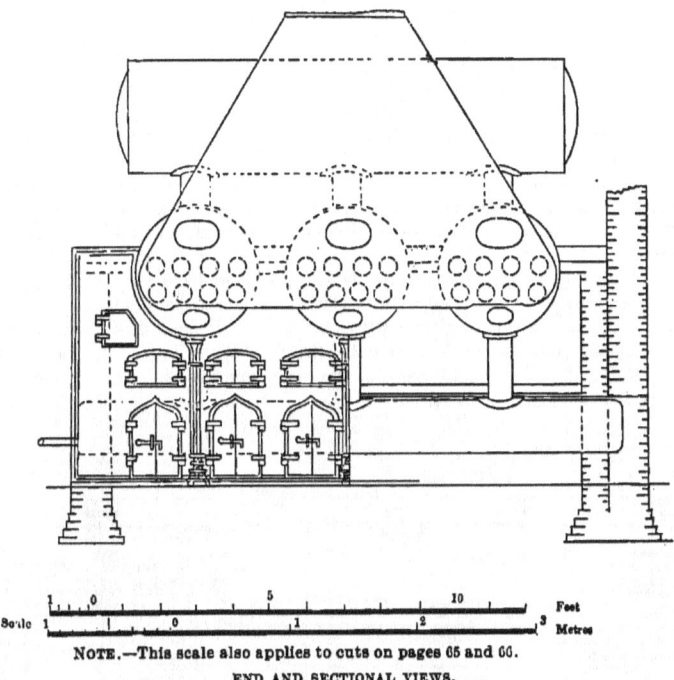

NOTE.—This scale also applies to cuts on pages 65 and 66.
END AND SECTIONAL VIEWS.

BOILERS AT WORKS OF ST. LOUIS LEAD & OIL CO., ST. LOUIS, MO.

yond which it projects enough to attach the bottom blow-off. There is neither dome nor mud leg. Steam is taken by a 6-inch sheet iron dry pipe. The heads are stiffened and braced in the manner shown. We suggest as an improvement that the dry pipe be taken from the top of the back head, and that the safety valve be connected thereto. By so doing, no opening need be made in the shell. All longitudinal seams are double-rivetted. The depth of furnace, 40 inches from shell to grate bar, has been much criticized, but it has answered very well.

All sorts of devices for smoke prevention, by the admission of air, etc., have been tried on this furnace, and the conclusion reached is that an intelligent fireman, with a moderate amount of work, will do more to prevent smoke than anything else; but that crowded as these boilers often are there is no way to prevent smoke.

The use of a central space for circulation is quite frequent, but it is usually not more than 5 or 6 inches. In most cases, however, the boiler is stuck as full of tubes as it can be, and they are placed so close to the shell that straining and grooving occur from the unequal expansion.

We illustrate, as conforming more closely to the usual practice, a 60-inch boiler designed by the Hartford Boiler Inspection and Insurance Company.

In some cases the products of combustion are taken over the shell after coming through the tubes. The use of domes upon the top of shell is almost universal, but there are objections to them which will be given later.

Very many compounds of these simple types have been made. The French boiler is often used with fire tubes in the upper shell, and a combination of two short cylinders, the lower and larger full of tubes, while the upper and smaller is used as a steam drum. This is introduced by the name of "compound boiler;" but as this name is used for many other forms, it is hardly distinctive.

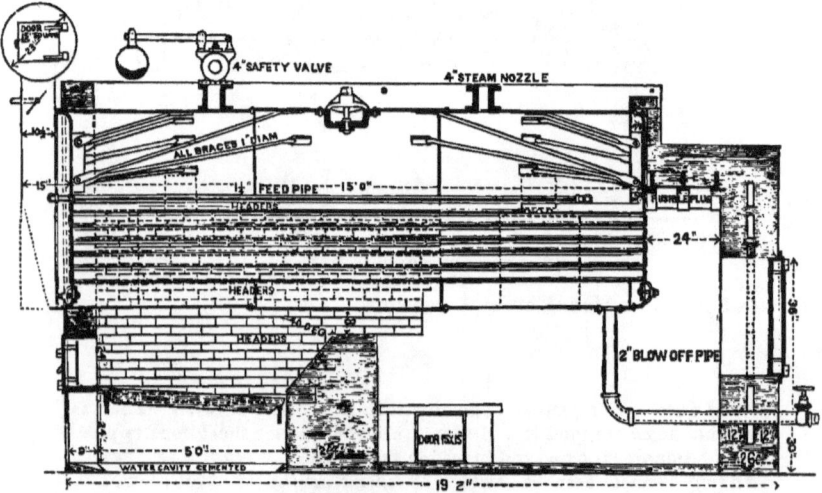

SECTIONAL ELEVATION.
60-INCH HORIZONTAL TUBULAR STEAM BOILER. DESIGNED BY HARTFORD STEAM BOILER INSPECTION & INSURANCE CO.

## SPECIFICATION

*For 60-inch Horizontal Tubular Steam Boiler, Prepared by The Hartford Steam Boiler Inspection and Insurance Company.*

| | |
|---|---|
| TYPE. | Boiler to be of the Horizontal Tubular type, with Overhanging Front and Doors complete. |

# EXTERNALLY FIRED STATIONARY BOILERS.

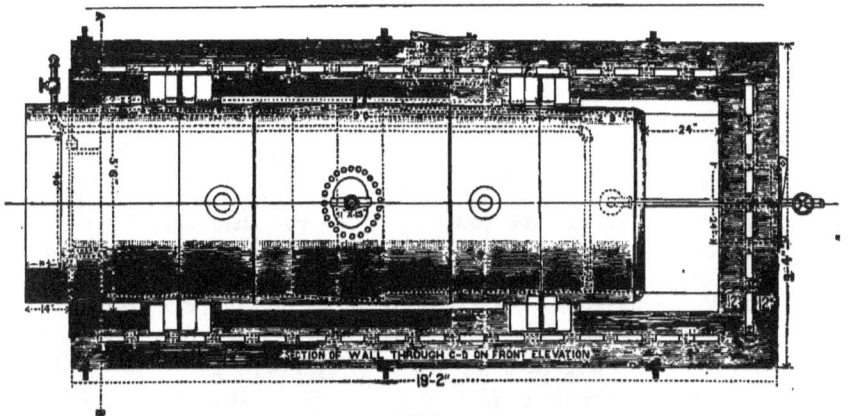

PLAN.

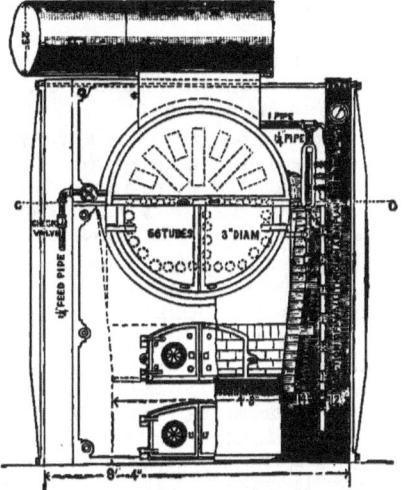

TRANSVERSE VIEWS.

| | |
|---|---|
| DIMENSIONS. | Boiler to be 16 ft. 3 in. long outside, and 60 in. in diameter. Tube heads to be 15 ft. apart outside. |
| TUBES. HOW SET AND FASTENED. | Boiler ...... to contain 66 best lap-welded tubes, 3 in. in diameter by 15 ft. long set in vertical and horizontal rows, with a space between them, vertically and horizontally, of not less than one inch (1"), except the central ver- |

tical space, which is to be two inches (2"), as shown in drawing. No tube to be nearer than 3 in. to shell of boiler. Holes through heads to be neatly chamfered off. All tubes to be set by a Dudgeon Expander, and slightly flared at the front end, but turned over or beaded down at back end.

### FOR IRON PLATES.

**QUALITY AND THICKNESS OF IRON PLATES.** Shell plates to be ⅜ of an inch thick, of the best C. H. No. 1 iron with brand, tensile strength, and name of maker, plainly stamped on each plate. Tensile strength to be not less than 50,000 lbs. per square inch of section, with a good percentage of ductility. Heads to be .... of an inch thick, of the best C. H. No. 1 Flange iron.

### FOR STEEL.

**STEEL PLATES.** Shell plates to be .... of an inch thick, of homogeneous steel of uniform quality, having a tensile strength of not less than 60,000 lbs. per square inch of section, nor more than 65,000 lbs, with 45 per cent. ductility, as indicated by the contraction of area at point of fracture under test. Name of maker, brand and tensile strength to be plainly stamped on each plate. Heads, to be of same quality as plates of shell in all particulars, .... of an inch thick.

**FLANGES.** All flanges to be turned in a neat manner to an internal radius of not less than two inches (2"), and to be clear of cracks, checks, or flaws.

**RIVETING.** Boiler to be riveted ¾-inch rivets throughout. All girth seams to be single riveted. All horizontal seams to be double staggered riveted. Rivet holes to be punched or drilled so as to come fair in construction. *No drift-pin to be used in construction of boiler.*

**BRACES.** There are to be twenty (20) braces in ..... boiler —ten (10) on each head, none of which are to be less than three (3) ft. long. Braces to be made of best round iron, of one (1) inch in diameter, and of single lengths.

**HOW SET AND FASTENED.** There are to be seven (7) lengths of T iron, four (4) inches broad and one-half (½) inch thick, four (4) being eight (8) inches long, two (2) being sixteen (16) inches long, and one (1) eighteen (18) inches long, placed radially; and riveted with ¾-inch rivets to each head, as shown in drawing. The holes for fastening the braces to these radial brace-bars are all to be drilled. The braces are to be fastened with suitable jaws and turned pins or bolts, so as to realize strength equal to inch round iron. Braces to be set as shown in drawing, and to bear uniform tension.

| | |
|---|---|
| MAN-HOLES. | Boiler....to have one man-hole, eleven (11) inches by fifteen (15) inches, with strong internal frame (as shown in drawing), and suitable plate, yoke, and bolt, the proportions of the whole such as will make it as strong as any other section of the shell of like area. |
| HAND-HOLES. | Boiler ...... to have two hand-holes, with suitable plates yokes, and bolts, located one in each head below the tubes, as shown in the drawing. |
| NOZZLES. | Boiler ...... to have two cast-iron nozzles, four (4) inches internal diameter, one for steam and the other for safety-valve connections, securely riveted to boiler. |
| WALL-PLATES. | Boiler ...... to have four cast-iron lugs, two on each side, the rear lugs each to rest on three transverse rollers, one inch in diameter, which are to rest on suitable cast-iron wall-plates, as shown in drawing, front lugs to rest on suitable wall-plates, without rollers. For blow-out connection, one plate, ½-inch thick, to be secured with rivets driven flush in inside of the shell, and tapped to receive a two (2) inch blow-pipe. |
| BLOW-OUT. | |
| FRONT. | Boiler ...... to be provided with cast-iron front and all the requisite doors and fastenings for facility of access to tubes, furnace, and ash-pit. |
| BUCKSTAVES. | Boiler ...... to be provided with ...... buckstaves; also all bolts, rods, nuts and washers, anchor-bolts to extend in setting beyond bridge-wall; also bearer and grate bars (pattern to be selected); also cast-iron door, to be at least two (2) feet by three (3) feet and provided with liner plate, for back tube door—and ...... door fifteen inches by fifteen inches for flue for side or rear end. |
| GRATE BARS. | |
| FITTINGS. | Boiler ...... to be provided with one safety-valve, ...... inches in diameter, .... inch steam gauge of standard make, three gauge cocks properly located, also one glass water gauge, a 2-inch open way blow-valve, and feed and check valves, each one and one-quarter inch. |
| FUSIBLE PLUG. | Boiler......to be provided with a fusible plug so located that its centre shall be two inches above upper row of tubes at back end. |
| DAMPER. | Boiler ...... to be provided with a damper with suitable hand attachments, easily accessible at the front of the boiler, damper to be fitted to the throat of the smoke-arch, as near as practicable to the tube openings, and of area equal to the cross section of all the tubes. |
| | The size and description of parts to conform substantially to the details of the accompanying plan. All the above to be delivered at.................. ................ |

and all the material and workmanship to be subjected to the inspection of and approved by the Hartford Steam Boiler Inspection and Insurance Company.

The axes of the cylinders are sometimes placed vertically as well as inclined. Of the water tube type we will only mention one in which a number of cone-shaped tubes hang from a horizontal drum into the fire, and Cadiats in which three horizontal drums are connected to a fourth above them as a steam drum, and from the former vertical cylinders of ten or twelve inches hang; those from the inner drum are connected at the bottom by a horizontal drum over the furnace, while those from the side drums are carried below the level of the grate, and each has hand hole and plate accessible from below. This boiler seems to combine many good features.

A form of upright fire tube externally fired boiler, introduced by Mr. George H. Corliss, of Providence, is shown by an example taken from the Pawtucket Water Works, at Pawtucket, R. I. These boilers when moderately worked with a low water have given steam in a superheated state.

There are three boilers at the Pawtucket Water Works, each four feet in diameter and fourteen feet high, carried on two cast-iron man-head trunnions. There are forty-eight 3-inch tubes arranged with a central gangway and a mud pan stayed to the bottom by stay bolts and tied to the top by rods. The shell is partly exposed to the fire and the circulation is up on the outside and down the middle, thus depositing sediment in the "mud pan." The steam connections are made by a casting acting as stiffener for the upper head. The water line may be varied from five to nine feet,—in the lower places about $20°$ superheating is obtained.

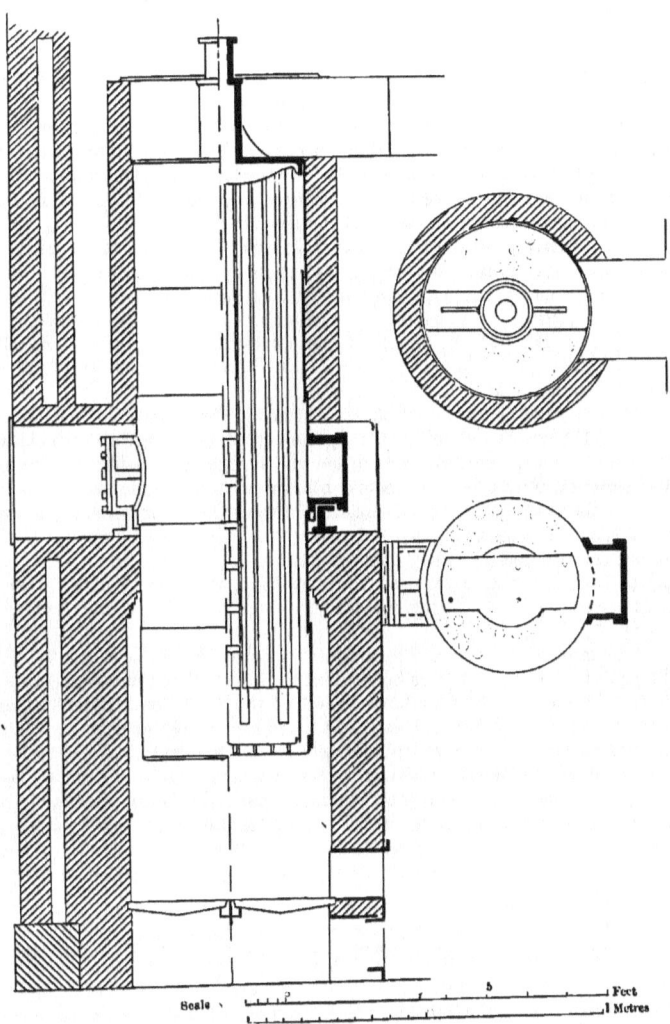

CORLISS BOILER AT THE PAWTUCKET WATER WORKS,
PAWTUCKET, R. I.

## CHAPTER IV.

### INTERNALLY FIRED STATIONARY BOILERS

As but very few horizontal internally fired boilers are used in the United States for stationary purposes, we have quoted entire, by the kind permission of the author, a paper by Mr. Lavington E. Fletcher, Chief Engineer of the Manchester Steam Users' Association, a boiler insurance company in England, as the highest authority on this subject in the world; and the paper is so admirably adapted to our purpose and so well written that we preferred to leave it in the original, fearing that it might be injured if incorporated with our work in any other manner.

The examples selected are: A single-flued internally fired boiler,— a Cornish boiler exhibited and tested at the Dusseldörf Exhibition. This boiler was constructed with the corrugated flue introduced by Fox, and therefore differs in this respect from the older forms of Cornish boilers, it is therefore adapted for much higher pressures than were formerly carried. The setting is not an example of ordinary practice, which is of the Lancashire type.

Double-flued boilers—a pair exhibited at the Vienna Exhibition of 1873, by Messrs. D. Adamson & Co., the introducers of the flange joint used in the flues. This boiler is rather shorter than that recommended by Mr. Fletcher, but in most respects conforms to his suggestions.

For the Galloway type we illustrate three out of four boilers built by the Edgemoor Iron Company for the Crystal Plate Glass Company: their works are situated 23 miles south of St. Louis. These boilers were put in in 1880, and while exposed to very hard work have given every satisfaction. The specifications under which they were built are appended.*

The most usual type of internally fired stationary boiler in the United States, is the locomotive, this being used by most of the railroad companies in their machine shops, and by several of the lately constructed water works, such as at Lawrence, Mass., and the Calumet and Hecla mines, Mich. The old-fashioned "drop return flue" or "tubular" boilers introduced by Mr. Kirkwood at the Brooklyn Water Works and copied at many places, are open to all the criticisms of Mr. Fletcher's paper, as being weakened by having large holes cut in the shell, rendering them unfit for the higher pressures now universal. At the Buffalo Water Works the boilers are of the North River type, as also those at the Cleveland Water Works—the latter being very large.

---

*The price of these boilers at Edgemoor, Delaware, on board the cars was $3,880 each.

The example selected is one of Mr. Leavitts, this being one of the latest. The vertical boiler with internal furnace is subject to almost endless modification, and we select for illustration a very good example designed by the Hartford Boiler Inspection and Insurance Company.

## ON THE LANCASHIRE BOILER.

### ITS CONSTRUCTION, EQUIPMENT, AND SETTING.

[A paper by MR. LAVINGTON E. FLETCHER, Chief Engineer of the Manchester Steam Users' Association, read before the Institution of Mechanical Engineers, London.]

The Lancashire type of boiler differs only from the Cornish in one point, namely, that the Lancashire boiler has two furnace tubes, whereas the Cornish has but one. In both types of boilers the shell is cylindrical, the ends are flat, and the furnace tubes are carried through from front to back, below the ordinary water line, while the boilers are laid horizontally and fired internally. Internal firing is essential either to a Lancashire or to a Cornish boiler. It is a mistake to speak of an internally fired Lancashire or an externally fired Cornish boiler, though this is frequently done. If the fires are taken out of the furnace tubes of a Lancashire boiler and put underneath, it is a Lancashire boiler no longer, but becomes an externally fired double flued boiler, and if a Cornish boiler be treated in the same way it becomes an externally fired single flued boiler. These boilers owe their names to the counties in which they were first brought into general use. The single furnace boiler was introduced early in the present century by Trevethick in Cornwall, and is therefore called Cornish. The double furnace boiler was introduced in 1844, by Fairbairn and Hetherington, in Manchester, and is therefore called Lancashire.

In laying down Lancashire boilers, the fact has been too frequently lost sight of, that directly a fire is lighted within them, they begin to move. The flat ends to breathe outward, the furnace tubes as well as the shell to hog upwards, and the whole structure to elongate. If sufficient allowance is not made for these movements, straining and sometimes rupture occurs; while tendency to this is frequently aggravated by putting in an extra thickness of metal with a view of adding strength, the additional thickness increasing the unequal expansion of the parts. For some years the writer has had opportunities of observing a large number of boilers of the Lancashire, as well as other, types in work under the inspection of the Manchester Steam Users' Association, and from these observations he has endeavored to mature as complete a boiler, in construction, equipment, and setting, as possible. In doing this the following points have been kept in view: to make the boiler safe for a working pressure of from 75 pounds to 100 pounds per square inch, to make the structure elastic so that it may not be rent by the movement of the parts consequent on alternate expansion and contraction, but may be able to endure the work of years; and to set the boiler and arrange the fittings so that the whole shall be above board and accessible for inspection. The writer does not agree with the view too generally held, but most obstructive to improvement, namely, that anything will do for a boiler, and that it is only a boiler after all. He thinks that a boiler should receive as much attention as an engine, that it should be made with as much accuracy and attended with as much care, that the fireman should not be condemned to work in a dark dirty hole called a stoke hole, and the boiler assumed to be black and grimy, but that the boiler should be placed in a suitable house kept bright and cheery, and the fittings as well as the whole structure kept clean and in first-rate working order; also that the fireman should be stimulated to become as proficient in the art of using his shovel and managing the fire, as a fitter in using file and erecting an engine. If this practice happily adopted by some were to become general, and first-class boilers were laid down, instead of low priced ones, the scientific boiler-maker would have fairer scope, the steam user would derive economy, and the public would be benefited by the prevention of explosions as well as by the abatement of the smoke nuisance.

The Lancashire boiler has many variations beside the simple form already described. There is the Galloway boiler, in which the furnaces instead of running through from one end to the other, unite in an oval flue strengthened by water pipes. There is the multi-tubular, in which the furnace tubes unite in a combustion chamber from which a number of small flue tubes about three inches in diameter and six feet long run to the back of the boiler. There is Hills multi-flued boiler, in which seven flues about 11 inches in diameter, and 8 to 10 feet long, take the place of the small tubes in the multi-tubular boiler. There are also others in which the furnace tubes branch off to the sides or bottom of the shell, instead of running right through to the back end. To all these variations in the Lancashire boiler and also to the Cornish, this paper applies as to the construction of the shell and furnace tubes, as well as in regard to the equipment and setting of the whole.

To assist in the construction of Lancashire boilers for high pressure, the Manchester Steam Users' Association authorized the construction of a boiler expressly for undergoing a series of hydraulic bursting tests, and the manufacture of the boiler was entrusted to Mr. Beeley, of Hyde Junction, who has heartily seconded the views of the Association and rendered valuable assistance in the prosecution of the trials.

This experimental boiler is seven feet diameter, which is the usual size for mill service. It is adapted for a working pressure of 75 pounds per square inch and its construction is as subsequently described: a number of experimental bursting tests have already been made, careful observations being taken of the behavior of the boiler under pressure. These tests have already furnished valuable information, and when completed will be fully reported. Some of the results are given in this paper. In order to preserve the precise form and character of the rents, the solid plating around them has been cut out intact: several of these specimens are exhibited at the meeting. An actual end plate of a boiler seven feet in diameter, equipped with the usual fittings, has also been prepared for exhibition, this boiler front has been made by Mr. Clayton, of Preston, who was one of the earliest to assist the writer in getting up a first-class boiler equipment for the Manchester Steam Users' Association.

## CONSTRUCTION.

*Dimensions.*—Short boilers are found to do more work, in proportion, than long ones. This has been confirmed by experiments on the rapidity of evaporation by Mr. Charles Wye Williams and others. Also short boilers strain less than long ones and are, therefore, less liable to need repair. A length of 30 feet should be the maximum, while with regard to the minimum some Lancashire boilers to suit particular positions have been made as short as 21 feet and found to work well though the fittings become rather crowded. The length recommended and now generally adopted is 27 feet.

The diameter of the boiler is governed by the size of the furnaces, which should not be less than 2 feet 9 inches, to admit of a suitable thickness of fire, and afford convenience in stoking. Thick fires are more economical than thin ones.

The space between the two furnace tubes should not be less than 5 inches and that between the furnace tubes and side of the shell 4 inches in order to afford convenient space for cleaning and for the free circulation of the water as well as to give sufficient width of end plate for enabling it to yield to the expansion and contraction of the furnace tubes. With this width of water space it will be found that furnace tubes having a diameter of 2 feet 9 inches require a shell of 7 feet, which will afford a headway of about 2 feet 9 inches from the crown of the furnaces to the crown of the shell. A furnace 3 feet in diameter gives room for a better fire than one 2 feet 9 inches, but it requires a shell 7 feet 6 inches in diameter. For high pressures the smaller diameter of 7 feet is generally preferred and has come to be adopted as a standard size for mill boilers throughout Lancashire, though one of 7 feet 6 inches makes a good boiler and gives a greater horse-power per foot of frontage than one of 7 feet diameter. The diameters, both of the shell and of the furnace tubes, are measured internally, that of the shell being taken at the inner ring of the plating.

*Ends.*—The ends, more especially the front, are the seat of the grooving action which occurs in Lancashire boilers when disproportioned. These grooves occur inside the boiler and around the furnace mouth. They are the product of mechanical and chemical action combined. The plate is fretted by being worked backwards and forwards by the movement of the furnace tubes, consequent on the action of the fire, and when in that condition is attacked by the acidity of the water. To prevent this grooving the ends should be rendered elastic so as to endure the buckling action without fatigue. To secure this elasticity there should be not only a sufficient width of end plate between the two furnace tubes as well as between them and the shell as already explained, but also a space of 9 inches between the centre of the bottom rivet in the gussets and those at the furnace mouth.

Also five gusset stays are found to work better than any other number. With five gussets one falls on the centre line, which is not only the weakest part of the front end plate, and thus where it requires the most support, but also where it can be held fast without resisting the movements of the furnace tubes. The part of the end plate that should be left free is immediately over the furnace crowns. With four gussets the end plate is more unguarded at the centre, which is the weakest part, and more bound immediately over the furnace tubes, which is the line of motion.

The thickness of the end plates is sometimes as much as $\frac{3}{4}$-inch for pressures of 60 pounds per square inch. This thickness, however, is quite unnecessary, and only tends, by its rigidity, to cramp the furnace tubes and strain the parts. Half an inch has been repeatedly and successfully adopted in boilers for pressures of 75 pounds per square inch and $\frac{9}{16}$ of an inch when that pressure has been exceeded. These thicknesses have proved amply sufficient. In applying the hydraulic tests to boilers of the construction and proportions now described before leaving the maker's yard, it is the practice to carry the pressures up to about 150 pounds per square inch, and to strain fine cords across the flat ends to act as straight edges from which to gauge the ends at twelve points, measurements being taken before the test, during the test, and after the test. It is found, as a rule, that the plate under pressure bulges outward at the centre from $\frac{1}{16}$ of an inch to $\frac{1}{8}$ of an inch, and on the removal of the pressure returns to its original position without suffering any permanent set. In the experimental hydraulic bursting tests the ends, though only $\frac{1}{2}$-inch thick, have stood a pressure of 275 pounds per square inch without leakage or any appearance of distress, but on the pressure being raised to 300 pounds the front end plate displayed signs of weakness in the vicinity of the mud-hole beneath the furnace tubes. With this exception the greatest bulging was $\frac{1}{8}$-inch at the front and $\frac{3}{16}$ of an inch at the back; while the greatest permanent set was only $\frac{1}{16}$ of an inch.

Longitudinal stays are frequently introduced to assist the end plates. In the experimental tests the longitudinal stays were taken out, so that it is clear that they are not absolutely necessary where the gussets are substantial. Should it, however, be thought desirable to adopt them, either as an assistance to the gussets when too weak, or as an extra precaution, they will be found easy of introduction. They are, therefore, shown in the diagrams, and it will be observed that they are secured at each end with double nuts, one inside the boiler and one outside, and one placed as much as 14 inches above the level of the furnace crowns, and as close together as convenience will allow. When placed directly over the furnace crowns and only a few inches above them they confine the furnace tubes too strictly and straining ensues. A single stay on the vertical centre line of the front end plate is correct in principle, but two are more convenient in application.

To increase the elasticity of the front end plate it is attached to the shell by an external angle iron ring rather than by an internal one, or by flanging. It is not necessary to attach the end plate at the back of the boiler with an external angle iron ring, and when this has been done, the angle iron has been found to be injured by the action of the flame. Both of the end plates, instead of being made in two pieces rivetted together at the joint, are welded so as to afford a flat surface, which in the case of the front plate is more convenient for the attachment of the mountings. Also both of them are turned in the lathe at the outer edge so as to be rendered perfectly circular and are bored out at the openings for the furnace tubes.

*Furnace Tubes.*—The longitudinal joints of the furnace tubes are welded when the plates are of iron, and double-rivetted when of steel, each belt of plating being made in one length and thus having but one longitudinal joint. All the transverse seams of rivets are strengthened by Adamson's flanged joint or with an encircling hoop, either of Bowling iron, T-iron, or other approved section. One of the evils that has attended internally fired boilers has been the frequent collapse of the furnace tubes, but this danger is completely avoided by strengthening the tubes as just described, whereby, instead of being weaker than the shell as before, they are rendered stronger.

This has been shown by the experimental bursting tests, in which, while the shell has been burst repeatedly, the furnace tubes have not suffered at all nor shown any movement on being gauged. In some cases Petrie's pockets, and in others Galloway's conical water pipes are introduced as a caution against collapse; while in others again the water pipes are made parallel, and either rivetted or welded in place so as to form one piece with the flue tube. In all cases, however, the transverse seams of rivets over the fire should be strengthened with flanged seams or encircling hoops, and it is considered desirable to continue this mode of construction throughout the entire length of the boiler, whether water pockets or water pipes are introduced or not. The thickness of plates in the furnace is sometimes as much as $\frac{1}{2}$ inch. This leads to violent straining and frequent leakage at the furnace mouths and other transverse seams of rivets. Many 2 ft. 9 in. furnace tubes, though only $\frac{5}{16}$-inch thick have stood a hydraulic test of 120 pounds per square inch without movement and have worked satisfactorily for years at a steam pressure of 60 pounds. It is advisable, however, to have them a little thicker than this, in order to afford a margin for waste through corrosion, and also when the flanged seam is adopted in order to allow for the thinning that occurs in drawing the metal to make the flange. A thickness of $\frac{3}{8}$ of an inch is sufficient for a working pressure of 75 pounds per square inch, $\frac{13}{32}$ for a pressure of 80 or 90 pounds, and $\frac{7}{16}$ for 100 pounds per square inch.

Stays are sometimes introduced for tying furnace tubes to the outer shells in order to support them. Such stays are, however, in the Lancashire boiler unnecessary, and when rigid, are decidedly objectionable;

furnace tubes should be left free to move. As soon as a fire is lighted within them, the top of the tube becomes hotter than the bottom and elongates. This makes the tube arch upwards. In conducting a series of trials in 1867 and 1868 for the South Lancashire and Cheshire Coal Association on the evaporative efficiency of their coals, and also on the comparative merits of different boilers, the writer had three gauge rods attached to the crown of the furnace tubes of two Lancashire boilers and carried up vertically through the external shell by means of brass stuffing boxes, so that a ready opportunity was afforded of witnessing the rise and fall of the furnace tubes, while as the gauge rods divided the tubes in equal lengths a comparison could be drawn as to the movements of the different parts.

Constant observation showed that the distortion of the tubes varied very much at different times, being most severe shortly after lighting the fires, while the colder the water to start with the greater was the rise of the crown. As soon as the water became generally heated the gauge rods retired to their old position, and the distortion of the furnace tubes seldom lasted more than an hour.

The boilers were 28 feet long, the furnace tubes of steel, $\frac{5}{16}$ of an inch thick in one case, and of iron, $\frac{3}{8}$ of an inch thick in the other. Care was taken not to strain the boiler by severe firing, steam being got up with the dampers only partially open, yet the furnace tubes rose $\frac{3}{8}$ of an inch when the flame passed round the boiler in the external brickwork flues in the ordinary way, and $\frac{1}{2}$ inch when they passed off direct to the chimney without heating the outer shell. The curve that the flue appears to assume is not a segment of a circle; the gauge rod at a quarter of the length of the boiler from the front showed in one case as high a rise as the rod placed midway in the length of the boiler, and in another case $\frac{1}{16}$ of an inch more. This is just what might be expected from the local action of the fire and accounts for the grooving action being far more severe at the front end of a boiler than at the back, and shows the importance of affording greater elasticity at that part. Furnace tubes lashed to the shell often tear themselves away from it in ordinary work, and the fractured stays rubbing against the shell leaves a witness of its movements, the amount of which frequently exceeds that just mentioned. In one case a furnace tube that had a stay tying it to the top of the shell was found to have crumpled up the stay and broken it by an upward thrust, showing how little need there had been for tying to keep the furnace tube from drooping.

*Shell.*—The shell, which is $\frac{7}{16}$ of an inch thick for a pressure of 75 pounds per square inch, and $\frac{9}{16}$ of an inch for a pressure of 100 pounds, is composed of plates about 3 feet wide, which are laid in not more than three lengths round the circumference, in order that the longitudinal seams may clear the brickwork seatings. The longitudinal seams are so arranged as to break joint, and avoid the centre line along the top and bottom of the boiler. In all the longitudinal rents obtained under the experimental hydraulic tests the plates bulged outwards at the middle of their width, and this action was observed to a slight extent before rupture, showing that the greatest strain, and thus the point of first fracture, occurred at or near the centre line of each plate. This would seem to show that breaking joint is of practical advantage, and that a boiler composed of wide plates is not so strong as one composed of narrow ones.

There is no steam dome. Steam domes are expensive, weaken the shell, and often give trouble from leakage at the base; added to this, they are inconvenient in carriage as well as in revolving a boiler on its seat, as it is sometimes desirable to do for repairs. They are also inconvenient in covering the boiler over, and in the great majority of cases, if not in every instance, they are perfectly useless.

To prevent priming an internal perforated pipe is adopted in place of the dome. Under hydraulic pressure a steam dome 3 feet in diameter $\frac{7}{16}$ of an inch thick, and the whole of the shell plate at its base cut away, so as to form an opening as large as itself, the flange at the base of the dome ripped, at a pressure of 150 pounds per square inch.

At a second trial, with a dome of the same diameter, and a portion only of the shell plate cut away, the dome strained so much round its base and caused such violent leakage that a pressure of more than 235 pounds could not be obtained. At a third trial, the steam dome having been removed and refixed with stouter rivet heads, so as to resist the upward strain that was induced, the flange on the bottom of the dome ripped on the centre line of the boiler, at a pressure of 260 pounds per square inch. In this instance the workmanship was all good and sound; but in some cases, where domes are attached with inferior reedy angle irons, the weakening effects of the domes must be much greater. Steam domes clearly establish a weak point in a shell, and are better avoided.

The manhole is guarded with a substantial, raised mouthpiece of wrought iron, welded into one piece, flanged at the bottom and attached to the boiler with a double row of rivets, the thickness of the upper flange being $\frac{7}{8}$ of an inch, and of the body $\frac{3}{4}$ of an inch. This has been found to stand a test of 300 pounds per square inch without the slightest indication of straining. A raised wrought iron manhole mouthpiece is exhibited. It is too frequently the practice not to strengthen manholes with any mouthpiece at all. Many explosions have arisen from this cause, rents starting in the first place from the unguarded manhole, and then extending all over the boiler. The loss of strength is owing not simply to the amount of metal cut away by the opening, but also to the action of the cover, which in unguarded manholes is internal. This internal cover bears on a narrow edge of plating all round, and is driven outward by the pressure of the steam, and also pulled in the same direction by the bolts in tightening the joint. In fact the cover acts as a sort of mandrel, which, being forcibly driven through the manhole, splits the boiler open. A heavy hydraulic test shows this action of the cover by curling the boiler plate up around the manhole. Added to this, the joint is apt to leak, and thus to induce corrosion and thin the plate, which not only reduces its strength, but leads to extra force being applied to tighten the joint—several explosions have occurred after the joint has been re-made. It has been the general practice, until recently, to make the raised mouthpieces of cast iron. This, however, is not wise for the high pressures now in use. A raised manhole mouthpiece having a clear opening of 16 inches, which is the usual size, involves a hole in the shell plate of about 20 inches in diameter. The plate in which this hole is cut, unless it be duly strengthened, becomes the weakest part of the boiler when the longitudinal seams are double riveted, the furnace tubes suitably strengthened with encircling rings, and the ends well stayed, so that the stability of the entire structure depends upon the mouthpiece; if that fails the whole structure fails. Under these circumstances it is evidently unwise to risk the safety of the boiler on a piece of cast iron. This view is confirmed by the behavior of cast iron manhole mouthpieces under hydraulic pressure. Several have failed under the ordinary test at the boiler-maker's yard, while at one of the experimental bursting tests a cast iron manhole mouthpiece, of substantial pattern, measuring 1¾ inch thick in the lower flange, and 1 inch in the body, rent at a pressure of 200 pounds per square inch, though the metal exhibited a good, sound fracture. This specimen is exhibited to the meeting. It would appear that under pressure there is considerable upward strain on the plates around the mouthpiece, and that while wrought iron mouthpieces are able to accommodate themselves to this without distress, cast iron ones are not. These tests have shown that wrought iron manhole mouthpieces are much superior to cast-iron, and that the sooner cast-iron ones are superseded by wrought-iron ones the better.

The mud-hole at the front of the boiler beneath the furnace tubes is also fitted with a substantial mouthpiece. This, in some cases is external like the manhole mouthpiece, and in others internal; the internal ones have the advantage of being less in the way. In either case the surfaces at the joint between the body of the mouth-piece and the cover are faced true so that the parts may be brought together, metal to metal.

The safety valve and steam stop valve are sometimes grouped upon the man-hole mouthpiece instead of being fixed direct to the shell. This is done in order to reduce the number of openings, on the principle that the fewer holes made in the boiler the better. This argument is plausible but fallacious; the man-hole makes the largest opening and therefore exerts the greatest weakening effect. The weakest link in a chain is the measure of strength of the whole, so that fixing the steam stop valve and safety valve directly to a boiler with suitable fitting blocks does not weaken it. Moreover for convenience in attaching the fittings, these group man-hole mouthpieces are made of cast-iron, which, as already explained, is objectionable. It is therefore recommended that man-hole mouthpieces should not be complicated by the addition of the safety valves or other fittings, but that each should be fixed direct to the shell independently of the others.

*Blocks for the Attachment of Fittings.*—In old fashioned practice the fittings were bolted directly to the cylindrical portion of the shell. This led to the wasting of the shell through leakage at the joints, so that it has long since been the practice to rivet short stand pipes to the cylindrical portion of the shell, and bolt the fittings thereto, the joint surface between the flanges being planed up true. These stand pipes, frequently termed "fitting blocks," are not only more convenient for the attachment of the fittings, but also being rivetted to the plate and made of substantial section strengthen the plate round the hole cut in the shell. They are, as a rule, made of cast-iron, but it becomes a question whether with the high pressures now in use they should not be made of wrought iron. At one of the experimental bursting tests, a fitting block for a 6-inch steam valve box was found to give way before any other part of the boiler at a pressure of 275 pounds per square inch, though the flange was $1\frac{3}{8}$-inches thick, the body $\frac{7}{8}$ of an inch, and the metal sound.

*Seams of Rivets.*—Those running longitudinally in the cylindrical shell are all double-rivetted with $\frac{3}{4}$-inch rivets spaced about $2\frac{1}{2}$ inches apart longitudinally and 2 inches diagonally. The remaining seams throughout the boiler are single-rivetted, only the rivets being spaced 2 inches apart. To double rivet the transverse seams adds but little if any strength to the boiler, though it increases its weight and cost. It would appear that the strain upon the transverse seams of rivets in a Lancashire boiler is over-estimated. In a plain cylindrical boiler without furnace tubes, the strain on the transverse seams of rivets is precisely half that on the longitudinal seams.

By the introduction of the furnace tubes not only is the longitudinal strength increased but at the same time the area of the ends upon which the steam acts, is diminished also. So that in the Lancashire boiler the strain on the transverse seams of rivets is less than half that on the longitudinal seams. The force of this reasoning, however, is sometimes disputed and tie rods are introduced to support the transverse seams of rivets in the shell. But in the hydraulic bursting tests with the tie rods removed, the longitudinal seams of rivets were found to fail in every case before the transverse seams which never showed the slightest signs of distress and scarcely leaked a single drop, while some of the longitudinal seams under severe pressure leaked profusely.

The riveting is done by machine in preference to hand in the cylindrical shell, in the furnace tubes and as far as practicable in the flat ends. In the experimental bursting tests the machine work proved much tighter than the hand work. The rivet holes in the angle irons, T-irons, and flanged seams are drilled; those in the plates are punched by most makers, though by some the holes are drilled throughout, and the practice of drilling is strongly advocated by them. In investigating an explosion that occurred at Blackburn, in 1874, the mean tensile strength in twelve tests of a solid plate was found to be 21.19 tons per square inch, and in four tests of a punched plate 20.17 tons, showing a loss by punching of 1.02 tons per square inch, or about 5 per cent. The question of drilling versus punching, and also of the pitch and diameter of rivets, is one that deserves further consideration, and it may be added that a boiler 7 feet in diameter, and made of plates $\frac{7}{16}$ inch in thickness, having the longitudinal seams double riveted with $\frac{3}{4}$-inch rivets spaced 3 inches apart longitudinally, instead of $2\frac{1}{2}$ inches as usual, was found tight at a hydraulic pressure of 120 pounds per square inch. The edges of the plates at the longitudinal seams of rivets are planed and caulked tightly inside as well as out, though in many cases caulking is superseded by fullering.

*Material.*—As a rule boilers made under the inspection of the Manchester Steam Users Association, are of iron in the shell, while steel plates are very frequently introduced in the furnace tubes for a length of 9 feet over the fire, and sometimes from one end of the boiler to the other. For the furnace tubes steel plates have been found to give great satisfaction, but a little suspicion has been entertained with regard to their use for shells, seeing that the plates are then in extension, and that a small flaw through brittleness might extend till it produced serious consequences. "Best best" plates from first-class makers are always recommended, more importance being attached to ductility than to their tensile strength. Brands, however, are uncertain, and it is thought desirable that a complete system of testing should be adopted, and that before a boiler is made one plate out of the set proposed to be used should be tested as a check, the investigation having special reference to ductility. Lowmoor rivets are recommended and are frequently used.

## EQUIPMENT.

*Fittings.*—The fittings are so arranged that all those requiring frequent access are immediately within reach of the attendant when standing in front of the boiler. The feed is introduced on one side of the front end plate about 4 inches above the level of the furnace crowns, an internal dispersing pipe being carried along inside the boiler for a length of about 12 feet, and perforated for the last 4 feet of its length. On the opposite side of the front end plate is fixed the scum tap, to which is connected a series of sediment catching troughs fixed inside the boiler. In the centre of the end plate are two glass water gauges, one acting as a check upon the other, a pointer being fixed to show the correct height at which the water should be kept. Immediately above the water gauges is a dial pressure gauge, and above that a dead weight safety valve. Thus, whenever the attendant opens the furnace doors to charge the fires he has the height of the water and the pressure of the steam directly before him. Under his feet is the blow-out tap and behind him the coal supply, so that everything is ready to hand. He has not to climb a ladder to reach the water gauges or ascertain the steam pressure, nor to mount on the top of the boiler in order to regulate the feed supply. A handle for regulating the damper is frequently brought to the boiler front. On the top of the boiler are two safety valves, one a dead weight valve of external pendulous construction, the other a low water valve. But convenience of manipulation is not the only reason for this arrangement of fittings, and if the feed be cold and be introduced near the bottom of the boiler it is apt to induce local contraction and thereby strain the transverse seams of rivets near the bottom of the shell, but when introduced near the surface of the water and passed through an internal perforated pipe it becomes dispersed before falling to the bottom. Further, although non-return valves may be introduced they will sometimes fail and allow the water to escape, whereby the furnace crowns become bare and over-heated. When the feed inlet is placed above the level of the furnace crowns it will be seen that they cannot be drained bare by the non-return valve, but when placed at the bottom of the boiler, the boiler may then be emptied by such an occurrence.

All the joint surfaces are planed up true and the parts brought together, metal to metal, the edges of the flanges are turned as well as the bolts, while the heads of the nuts are shaped and the whole got up like a piece of engine work as shown by the end plate and fittings exhibited.

*Safety Valves.*—The dead weight valve which is of the Cowlum type is extremely simple and efficient. The centre of gravity of the load being below the seating, renders unnecessary either wing or fang for keeping the valve in position, and it has therefore no frictional surface to get tight or stick fast. These valves are loaded with flat annular plates or rings, and the shell is cast with mouldings around it at the bottom, which present the same appearance, the whole being so adjusted that each moulding as well as each annular plate represents a pressure of 5 pounds per square inch on the valve. Large numbers of these valves are in use, and they are highly approved. The diameter generally adopted is 4 inches, which requires approximately a load of 8 cwt. for a blowing-off pressure of 75 pounds, and 11 cwt. for 100 pounds per square inch. On this valve the addition of two or three bricks produces no appreciable effect, whereas at the end of a long lever the result would be different. To double the blowing-off pressure it would be necessary to add about 8 cwt. to the load for 75 pounds, and 11 cwt. for 100 pounds. Such an addition there would be great difficulty in attaching to the valve, and if it were done it would be so conspicuous as at once to call attention to the fact. The great weight required to load this valve is considered therefore to be a safeguard, and several explosions due to overloading have been met with which would have been prevented by its use.

The low water safety valve shown is of the Hopkinson type, but there are also the Kay and the Lloyd low water valves, which, though varying in detail are similar in their object. Each has a lever inside the boiler to which is attached a float so that when the water falls below the desired level the float falls also, and thus raises the valve and allows the steam to blow off, thereby not only giving an alarm but also lowering the pressure. Hopkinson's valve is a compound one, having one valve seated on another; the central portion is loaded by a dead weight inside the boiler and operated on by the lever in the event of low water, while the annular portion is loaded by an external lever and weight and lifts, along with the central portion in the event of high steam. These valves therefore blow off on the occurrence either of high steam or low water. The outer valve is 5 inches in diameter, and the inner one 2½ inches. The freedom of the steam valve can be tested by placing the hand on the lever when the steam is up, while the freedom of the low water apparatus can be tested by opening the blow-out tap and lowering the water level to within about 6 inches of the furnace crowns. To overload this valve without increasing the weight outside would necessitate getting inside the boiler and wedging down the dead weight. Under such circumstances the application of the hand to the external lever, when steam was up, would at once show that something was wrong, and even if this were not detected the external dead weight valve if free would come to the rescue while it would be seen at a glance if this were overloaded. It is sometimes recommended to have safety valves under lock and key, but it is preferred by the writer to have them thoroughly open, so that their publicity may be their protection. While it is fully admitted that no arrangement of safety valves can be constructed which cannot be tampered with by skilled malice, it is thought that the combination of the two valves just described forms a very safe arrangement.

*Furnace Mountings.*—The furnace mouthpieces are of wrought-iron, finished off with a neat brass beading and kept within the circle of the rivets so as to leave these exposed to view. The fire doors are fitted with a sliding ventilating grid on the outside and a perforated box baffle plate on the inside, the aggregate area of the air passages being about 50 square inches for arch door or about 3 square inches per square foot of fire grate. The fire grate is 6 ft long with three bars in three equal lengths about $\frac{3}{4}$-inch thick and spaced $\frac{3}{8}$ of an inch apart for windage. The bearers consist of two wrought-iron bars carried on wrought-iron brackets riveted to the sides of the furnace tubes. The standard length of grate is 6 feet, but a shorter one is productive of economy, though the concentration of the fire is more trying to the boiler and has been found, where the feed water has not been good, to injure the furnace plates and render it necessary to lengthen the grates.

*Brickwork and Flues.*—The boiler is set on side walls and rests on firebrick setting blocks, presenting a bearing surface 5 inches wide. The side flues are 6 inches wide at the top carried up to the level of the furnace crowns or a few inches above and down to the level of the bottom of the shell. The bottom flue has a width equal to the radius of the boiler and a depth of about 2 feet. The dimensions admit of ample room for inspection. By keeping the width of the bottom flue equal to the radius of the boiler the angle that the bearing surface of the seating blocks makes with the horizon is 30° for any diameter of shell.

The flame immediately after leaving the furnace tubes passes under the bottom of the boiler and returns to the chimney along the side flues. This is not the course approved by Mr. Pole in his treatise on the Cornish Pumping Engine published in "Tredgold on the Steam Engine," in 1844, in which the setting of the Cornish boiler is spoken of as follows:

"The heated current first impinges on the top of the tube over which "the highest and therefore the hottest portion of the water is lying, it then "passes along the side flues, where it finds the surfaces cooler than before, "and last of all it traverses under the bottom of the boiler where the cold- "est water will always be. By this means the fire current as it gradually "cools, is likewise gradually brought to act upon cooler water and thereby "the best opportunity is afforded for the extraction of the free caloric it "contains. The descending motion of the fire current as it cools in the "flues of the Cornish boiler is upon statical principles much more natural "and more calculated to prevent the unnecessary discharge of heat into the "chimney than the ascending principle of the ordinary boilers."

Allowing the last heat, however, to travel under the shell does not promote the circulation of the water, or at all events but slowly, so that in getting up steam the top of the boiler becomes hotter than the bottom from which straining ensues. If in addition to this the feed water when cold be pumped in or near the bottom of the boiler the straining at the transverse seams of rivets is intensified. Possibly the Lancashire boiler is more subject to straining and seam rending at the bottom of the shell than the Cornish as there is a greater body of dead water lying there in the Lancashire boiler, in addition to which the rate of combustion per square foot of fire grate is much more rapid in the Lancashire district than that generally adopted in Cornwall. In consequence of seam rents occurring at the bottom of Lancashire boilers when the last heat is carried underneath, the plan of passing the flame under the bottom immediately, on leaving the furnace tubes, and also of introducing the feed water near the surface has become the general practice.

The question of economy is met by the use of feed water heaters consisting of a number of water pipes placed in the main flue between the boiler and the chimney and kept free from soot by an automatic scraper. A good feed water heater will raise the temperature of the water to about 240°. This answers two good purposes; it economizes the waste heat escaping to the chimney and thus reduces the coal consumption, while at the same time it prevents local cooling, thereby preventing straining and saving repairs. It has been found by experiment that passing the flames from the furnace tubes around the outer shell instead of direct to the chimney adds but little to the yield of steam, though it promotes economy of fuel, at the same time that it keeps the boiler at a more equable temperature throughout.

The flooring or hearth plates at the front of boiler are set so as not to butt against the boiler, which is too often the case, but so as to be entirely below it, thus leaving the whole of the front end plate open to view. Where there is a range of boilers these flooring plates extend throughout the width of the boiler house and being finished off with a fender flange where abutting against the boundary walls of the building, as well as against the face of the brickwork setting, they present a very neat appearance. These plates are carried on a complete system of framing and are arranged for easy lifting. The hearth-pit beneath them is open from one side of the boiler house to the other, and in this is laid the main feed pipe as well as the discharge pipe from the blow-out and scum. This pit is about 3 feet wide by 2½ feet deep so as to afford room for access: the flue doors open into it. The face of the brickwork at the front of the boiler is set back 6 inches so as to leave the angle iron with its circle of rivets perfectly open. The front cross wall beneath the boiler is recessed around the blow-out elbow pipe so that it may be free to move should settlement of the boiler take place.

*Boiler Covering.*—The boiler is covered with an arch of brickwork, leaving a space of about 2 inches between it and the plates, and a layer of cork shavings or a coating of good boiler composition or other suitable non-conducting substance is introduced into this space. Openings finished off with bull-nosed bricks are worked round the fittings, so as to leave the ring of rivets by which they are attached to the shell exposed to view. Sometimes the boiler is covered simply with a layer of composition which should not be carried over the flanges of the fittings, as is too often the case, but should be stopped off by means of kerb hoops dropped around the flanges and a kerb cast-iron nosing to guard the front angle iron.

*Connections.*—All connections to boilers should be elastic so as to allow of their movement. If the main steam pipe be carried across the boilers and bolted direct to the junction valve, the joints are strained by the raising and falling of the boilers as they are set to work and laid off. To prevent this a springing length should be introduced between the steam stop valve and main steam pipe. Where the main steam pipe has a considerable length to travel to the engine, it should not be taken in a direct line, but should be carried round the boiler house or be led in a horse shoe shaped course to give elasticity; this is better than introducing an expansion joint which is not reliable. Sometimes expansion diaphragms are adopted, but these, when as much as 4 feet in diameter, have been known to lead to the fractures they were intended to prevent, the internal pressure causing them to bulge outward when it was expected they would allow the pipes to expand and thrust them inwards. A case of this sort has recently come under the knowledge of the writer in which the main junction valve was broken off by the thrust occasioned by the bulging of the expansion diaphragm. It is equally important that the feed connection should be elastic, and from the want of elasticity feed valve boxes have been known to fracture. For this purpose a copper elbow connecting pipe is introduced between the main feed pipe and the stand pipe; in some cases a wrought-iron horseshoe-shaped pipe has been adopted instead with very satisfactory results.

Connections between the steam stop valves and main steam pipe are frequently made to incline upwards, so that the water may drain back to the boilers. This plan is, however, objectionable, for when one of the boilers in a range is laid off, the connecting length becomes filled with water from condensation of the steam, which, cooling by radiation sets up a violent conflict with the steam whereby the pipes are sometimes fractured. The action may be illustrated by the commotion which occurs within a locomotive tender when the steam from the boiler is turned into it. Further than this, on opening the steam stop valve of a boiler that has been laid off the water lying on the top of the valve is apt to be carried forward by the rush of the steam like a water hammer, and sometimes to burst the pipe. To prevent this the steam pipe should drain towards the engine and not towards the boiler, its course being intercepted by a separator fixed as near the engine as convenient. The principle on which these separators act is that of making the steam take a sharp turn to shoot off the water mixed with it into a catch chamber prepared for the purpose. Many of these separators are now at work; the principle was advocated by Dr. Haycraft, five and twenty years ago.

*Weight and Cost.*—The weight and cost of such a boiler as has now been described is about 12 tons without fittings; with fittings 15½ tons. The cost at the present time (1876) delivered on the premises of the purchaser within a few miles of Manchester and including the attachment of the fittings is about £425. The plan of buying a boiler at so much per ton and then the fittings at so much extra, is quite given up in favor of purchasing the whole for one sum.

*Heating Surface.*—Such a boiler has heating surface in the external shell of 370 square feet; in the furnace tubes, without water pipes, 450 square feet, in the water pipes 30 square feet, making a total of 850 square feet. The fire grate has an area of 33 square feet. This gives for every square foot of fire grate 26 square feet of heating surface. The surface in feed water heaters varies; 60 pipes, each affording a heating surface of about 10 square feet are now frequently introduced per boiler, making a total heating surface of 600 square feet, or about three-fourths of that in the boiler.

*Working Results.*—Such a boiler as that described will burn without distress to the boiler from 15 to 20 tons of coal in week of 60 working hours, or from 17 pounds to 23 pounds per square foot of fire grate per hour. This may be done without making smoke. All that is needed is to maintain a good thickness of fire, throw on the coal little and often, admit a little air above the bars for a short time after firing, and avoid the use of the rake. The coal may either be spread over the whole surface of the fire, or thrown at alternate firing, first to one side of the furnace and then to the other on the "side firing" system introduced by Mr. Charles Wye Williams.

A Lancashire boiler experimented on at Wigan with furnaces 2 feet 7 inches in diameter and a fire grate 4 feet long evaporated 83.54 cubic feet of water per hour from a temperature of 100° at the rate of 10.44 pounds of water per pound of coal when burning 24 pounds of coal per square foot of fire grate per hour. With a fire grate 6 feet long, it evaporated 98.58 cubic feet of water at the rate of 10.37 pounds of water per pound of coal and burnt 19 pounds of coal per square foot of fire grate per hour. These results were obtained at atmospheric pressure with the help of a water heater with good round coal and without making smoke. The boiler described in this paper having furnaces 2 feet 9 inches in diameter would evaporate a larger quantity of water per hour. Such a boiler is found in practice to be capable, provided the steam be applied to a fairly economical engine, of developing 200 indicated horse-power per hour, and 20 indicated horse-power per lineal foot of boiler frontage, side flues included. A Cornish boiler under similar conditions is capable of developing 16 indicated horse-power per lineal foot of boiler frontage. This leads to a question of the utmost importance, namely, the one which the late Mr. Robert Stephenson defined as the "administration of the steam" and fuller information is yet needed as to the comparative advantage of working steam on the compound or single cylinder principle, also as to the value of steam jackets as well as with regard to the initial and terminal pressures most conducive to economy. These inquiries, though full of interest, cannot be entered upon in the present paper, but one of the essentials to economy is the power of raising high pressure steam steadily and safely, and this may be accomplished by the use of the Lancashire boiler.

The example taken for a Cornish boiler is one exhibited at Dusseldorf in 1880, and subjected to the competitive trial held there. It is 7 feet 2½ inches in diameter and 31 feet 1 inch long. The corrugated flue is 4 feet 3 inches in diameter. The corrugations are probably exaggerated as the usual practice is $6'' \times 1\frac{1}{2}''$ and is nearly 10 inches from the shell. The setting is peculiar in that the products of combustion after leaving the flue pass under the boiler and then return on the sides and top, thus giving dry steam. The pit at the rear end is to hold ashes which are blown by a steam jet out of the flue while in steam. The usual setting for the Cornish boiler is similar to the Lancashire.

This boiler gave the best results as to evaporative economy of all tried at Dusseldorf.

The proportions of this boiler are as follows:

| | |
|---|---|
| Grate area at trial | 13.8 sq. ft. |
| Grate area usual | 24.5 " |
| Heating surface | 841.5 " |
| Superheating surface | 193 " |
| Area over bridge | 4.63 " |
| Area through flue | 13.00 " |
| Area through side flue | 8.20 " |
| Area through damper at trial | 4.4 " |
| Ratio heating surface to grate | 53.2 " |
| Ratio grate to air space in grate | 4.31 " |
| Water space | 634 cu. ft. |
| Steam space | 259 " |
| Weight of boiler | 36736 lbs. |
| Brick setting | 1303 cu. ft. |

The evaporation reached at the trial was 10.854 pounds of steam from 1 pound of combustible at 75 pounds per square inch, with feed at 130° F. and a rate of 3.85 pounds of water per square foot of heating surface per hour. The side flues were then omitted and the boiler evaporated 5.92 pounds per square foot of flue surface and gave an evaporation of 8.17 per pound of combustible. The heating surface was then 24.6 times the grate.

The same makers are now building this boiler with the centre of the flue not in the same vertical plane as the centre of the boiler; they claim thereby an improved performance due to better circulation of the water.

The Lancashire boilers require no explanation after Mr. Fletcher's paper.

The fire-box boiler at the mines of the Calumet and Hecla Mining Co.,

# INTERNALLY FIRED STATIONARY BOILERS.

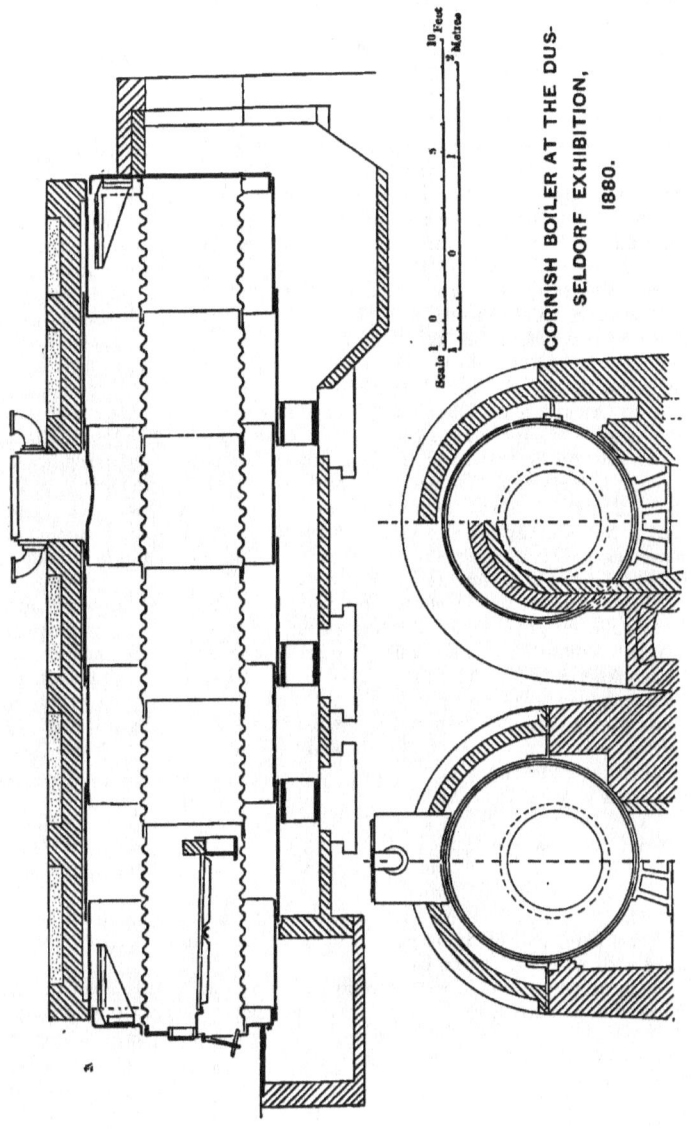

CORNISH BOILER AT THE DUSSELDORF EXHIBITION, 1880.

built from the designs of Mr. E. D. Leavitt, is probably the most expensive stationary boiler made, but it gives a very high economic evaporation when doing a good deal of work, and for steady work, night and day, is probably a good investment.

The shell, fire-box and combustion chamber are of the best quality Siemens-Martin steel, which possesses an ultimate strength of about 65,000 pounds, an elastic strength of about 49,000 pounds, an extension of $22\frac{9}{16}$ per cent. with a reduction of area at fracture of 58 per cent. The metal is therefore of excellent quality. The stays and braces of the "Best Ulster" iron. The iron rivets are "Burden's Best." The rivet holes are punched. The tubes are iron $3\frac{1}{4}$ inches in external diameter, and $4\frac{1}{2}$ inches from centre to centre, and 18 feet 1 inch long; the tube sheets are $\frac{1}{2}$-inch thick. The inside fire-box sheets and the combustion chamber are $\frac{5}{16}$-inch, the external fire-box $\frac{7}{16}$, and the shell $\frac{7}{16}$ of an inch thick. The shell is 7 feet in internal diameter, and is butt-jointed, with straps inside $\frac{3}{8}$ of an inch, and outside of $\frac{9}{16}$ of an inch in thickness. The junction of shell and fire-box is strengthened by inside and outside sheet on the ring of throat plates. The transverse seams are double-rivetted to the butt straps and over the fire-box the sheets are lapped. The longitudinal seams are treble-rivetted to the butt straps, the outer row having fewer rivets. The rivets in the fire-box are $\frac{3}{4}$ of an inch, in the shell $1\frac{3}{8}$ of an inch. The stay bolts are $1\frac{3}{8}$ of an inch in diameter and are spaced $4\frac{1}{2}$ inches centres horizontally and $\frac{3}{8}$ of an inch vertically. The fire-box is double, a water leg and two combustion chambers leading into a single chamber from which the tubes lead to a smoke-box.

Steam is taken by two slotted dry pipes through an 8-inch nozzle, and two 5-inch weight and lever safety valves are placed on a second 8-inch nozzle. A man-head on a nozzle gives access above the tubes, and one in the smoke-box below the tubes. The crown of the furnace and combustion chambers are slung by stays in a peculiar manner, adopted to give access through them for inspection; the heads are tied by light tie rods $1\frac{7}{8}$ inches round iron with swelled ends, and held by nuts and washers. Feed is taken through the top of the shell by a $1\frac{1}{2}$-inch brass pipe led through the water to the side of the boiler.

The fire-box rests on a cast-iron ashpit and the shell is carried on three adjustable cast-iron stands resting on balls. Two boilers are connected by a steam drum 24 inches in diameter and about 16 feet long,—the boilers being set 14 feet centres.

The front head is tied to the first sheet of the shell around the tubes by short bars, and the upper portion of both front and back head is stiffened with angle and T-iron bars.

The boiler is covered with a coating of plaster of Paris and sawdust $2\frac{1}{2}$ inches thick, covered with 1 inch of the best hair felt and a painted canvass cover. The coal used is an inferior quality from Ohio, being in evaporative value about 70 per cent. of the best steam coal. It is intended to use an artificial draft when desired, although the chimney is 150 feet high, and by this means to burn up to 40 pounds of coal to the square

# INTERNALLY FIRED STATIONARY BOILERS. 93

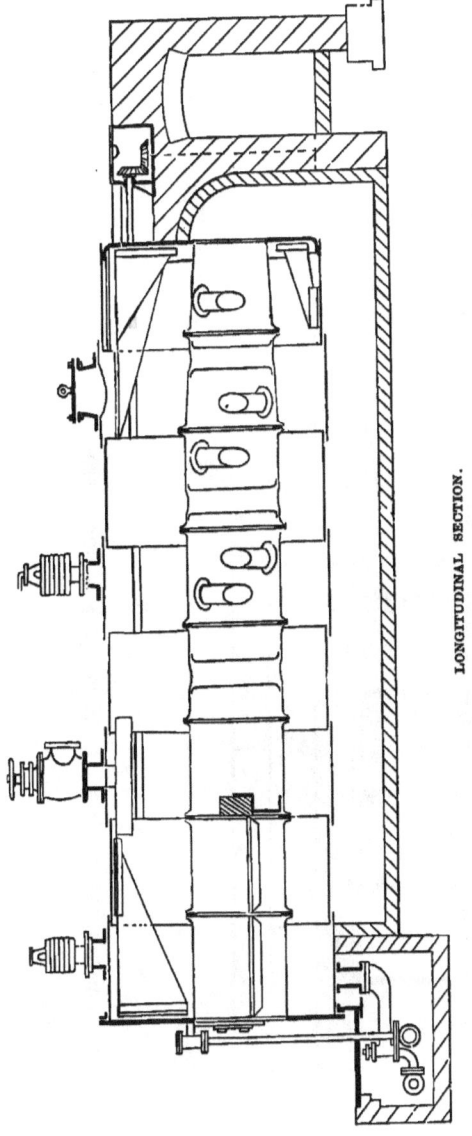

LANCASHIRE BOILER AT THE VIENNA EXHIBITION, 1873.

LONGITUDINAL SECTION.

94 STEAM MAKING; OR, BOILER PRACTICE.

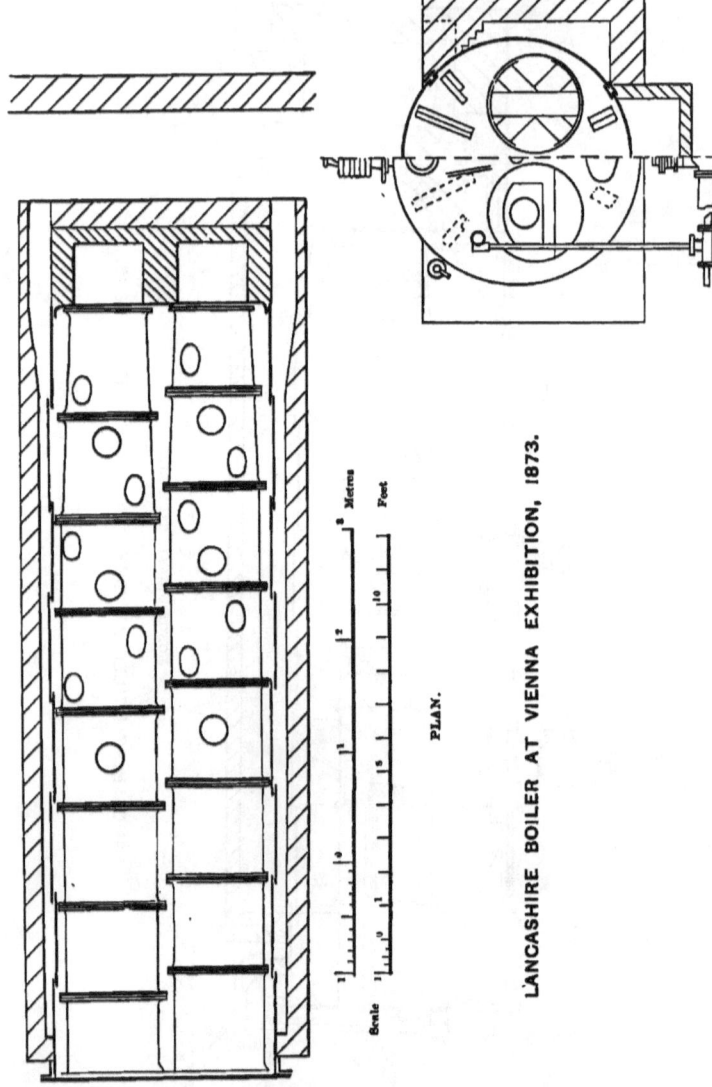

LANCASHIRE BOILER AT VIENNA EXHIBITION, 1873.

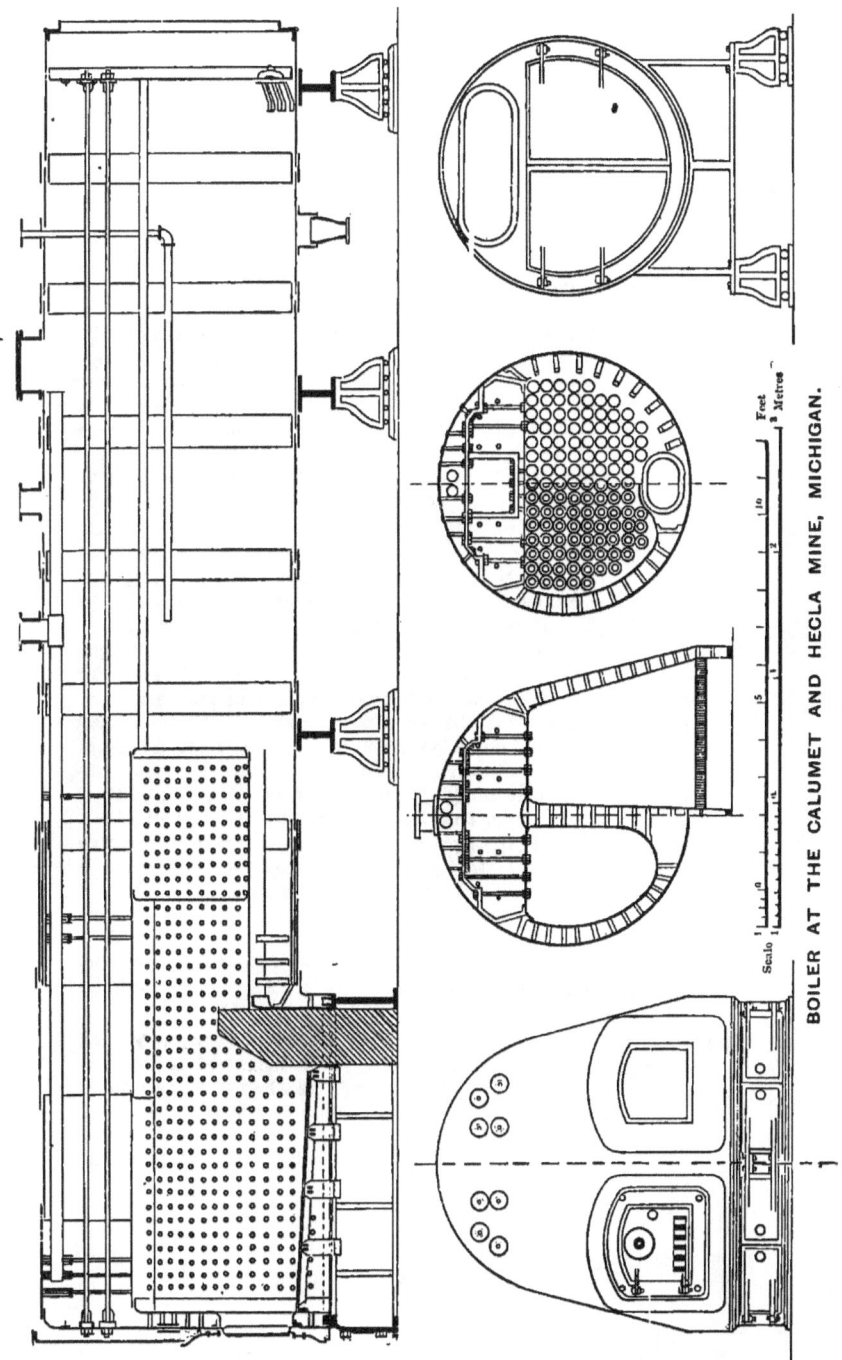

BOILER AT THE CALUMET AND HECLA MINE, MICHIGAN.

foot of grate per hour. The evaporation of 8.75 pounds of water from and at 212° per pound of inferior coal is maintained regularly, and a similar pair of boilers at the Lawrence Water Works gave 12.46 evaporation from Cumberland bituminous coal.

## SPECIFICATION

*For Galloway Boiler (1876 Patent), Edgemoor Iron Company. To Evaporate 50 Cubic Feet of Water per Hour. For Crystal Plate Glass Company, Crystal City, Mo.*

| | |
|---|---|
| SHELLS. | The shell to be 28 feet long by 7 feet diameter, and to be made of best cold blast charcoal flange-plates ½-inch thick. The longitudinal seams to be double-rivetted and to be crossed one-half the length of the plate so as to avoid a continuous line of rivets. The edges of all the plates to be planed and fullered, not caulked. Front end of shell to be provided with a solid welded ring of angle-iron 3 inches by 3 inches by ½-inch for attachment of the end. |
| FLUES. | The flues to consist of two furnaces each 2 feet 9½ inches in diameter formed of solid-welded rings of best cold blast charcoal fire-box plates ⅜ of an inch thick. The transverse seams to be formed by flanging the plates and inserting a solid welded ring between the two flanges. The two furnaces to unite behind the fire bridges into one flue of best cold blast charcoal flange plates made in accordance with above-mentioned patents. The flue being hollowed on its lower side to give more room for cleaning and examination, the necessary strength being obtained by bringing the tubes nearer together at their lower ends, and thus avoiding the necessity of objectionably thick plates which would otherwise be required for this part. This flue to be supported by means of 33 patent 'Galloway Cone Tubes having a diameter of 10½ inches at the top and 5½ inches at the bottom, the whole of these tubes to be interchangeable and to have all their flanges square to the centre line of the tubes, thus putting less strain upon the iron in their manufacture and thereby allowing a better job to be made. |
| END-PLATES. | The boiler ends each to be made in one piece $\frac{9}{16}$ of an inch thick, of the best cold blast charcoal flange iron. The front plate to be securely attached to angle iron of shell and the back end plate to be flanged. These end plates to be efficiently stayed by means of suitable gusset plates, which shall be fastened by double angle-iron to |

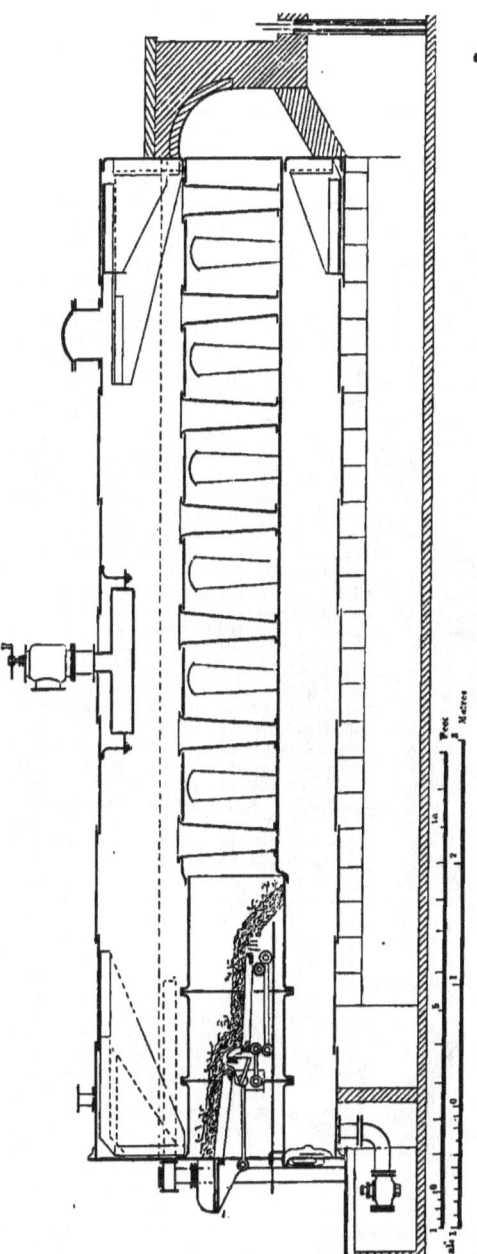

LONGITUDINAL SECTION.

GALLOWAY BOILERS AT CRYSTAL CITY, MO.

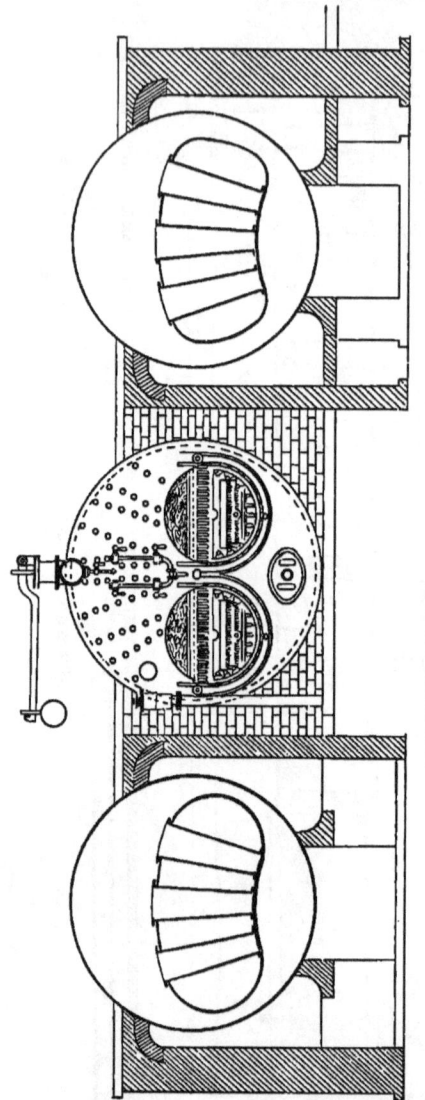

TRANSVERSE SECTION AND END ELEVATION.

GALLOWAY BOILERS AT CRYSTAL CITY, MO.

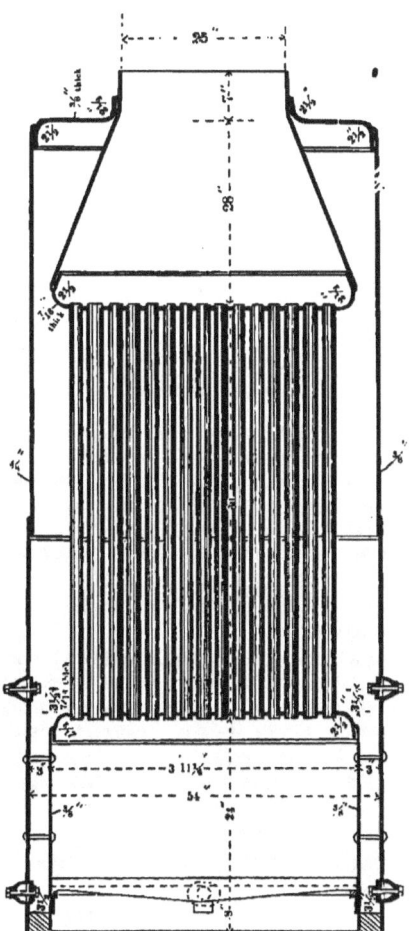

SECTIONAL ELEVATION.

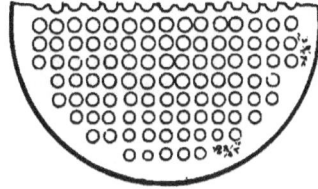

UPRIGHT BOILER,

Designed by the Hartford Steam Boiler Inspection and Insurance Company, Hartford, Conn. 10 feet long, 54 inches diameter. Heating surface: Shell, 4 square feet; Tubes, 419 square feet. Estimated horse-power, 30.

| | |
|---|---|
| | shell of boiler. These stays shall not be brought down too near the top of furnaces, but sufficient space shall be allowed for expansion. |
| Man-holes. | One wrought-iron manhole of large size shall be fixed on the top of boiler, and one of smaller size fixed on the front end plate below the flues, both to be rivetted on and to be faced across the whole surface of their flanges; cast-iron covers for the same to be provided with suitable bolts. |
| Testing. | Before leaving the works the boiler shall be tested with water pressure of 100 pounds per square inch, and a certificate of such test having been made shall be furnished. |

## MOUNTINGS FOR BOILER.

| | |
|---|---|
| Furnace fittings. | A set of suitable fire-frames and doors to be fitted on the front end of the boiler and each door shall be provided with a sliding shutter for the admission of a proper quantity of air for the prevention of smoke, also cast-iron hearth plates, bearing bars, fire bars of suitable length, cast-iron damper and frame, and ash-pit frame and plates. |
| Fusible Plugs. | An approved fusible plug shall be placed on the top of each furnace. |
| Blow-off Cock | One blow-off cock to be supplied with solid bottom and packed gland, having a flange at each end; also a suitable elbow pipe to be furnished for attaching the cock to the block previously rivetted on to the boiler. |
| Feed Valve. | A 2½ inch check feed valve shall be provided, the valve to be loose from the spindle so as to act as a check or non-return valve and to be set down by means of a hand wheel and screw. |
| Safety Valve. | Two 4-inch safety valves of the most approved construction. |
| Steam Nozzle. | One 7-inch junction valve fitted up with gun-metal valve and seatings, packed gland, hand wheel, etc. |
| Anti-Priming Pipe. | A cast-iron anti-priming pipe to be fixed inside the boiler and attached to the lower end of the above valve. |
| Water Gauge. | One set of brass fittings for duplex glass water-gauge with two glass tubes for the same. |
| Steam Gauge. | One 6-inch steam pressure gauge of best construction. |

## CHAPTER V.

### INTERNALLY FIRED PORTABLE, LOCOMOTIVE AND MARINE BOILERS.

Internally fired upright tubular boilers are not often used for stationary work unless the water is exceptionally good, on account of the difficulty of cleaning and examination, and the great liability to form scale or to corrode near the water line, and they will foam when crowded. Many devices of submerged smoke chambers, thus reducing the length of the tubes and, of course, reducing the heating surface, have been tried, and we find a good example in the upright tubular boiler recommended by the Hartford Boiler Insurance Company shown in the last chapter.

Among portable boilers, and classing with them the boilers for launches and fire engines, we have many varieties. The tubular upright with cylindrical furnace, or the horizontal with rectangular furnace, are the most usual. One of the simplest boilers ever made is a cylindrical shell with a conical internal furnace, with the upper head flanged to shell and upper portion of cone; the stack is attached directly to the top of the cone, which of course, is truncated. This boiler was used by the Lane & Bodley Company, of Cincinnati, Ohio.

The locomotive boiler has a cylindrical shell continued beyond the front head, making a smoke box, and with a rectangular furnace of considerable depth, surrounded on all but the bottom side by a second box with four flat sides and a semi-circular shell top; the sides are tied together by stay bolts, and the top held up by cross bars from which it is slung, or by stays from the shell above, the latter practice coming into favor with bad water as promoting a better circulation of water and depositing less scale.

We select for our illustrations the boiler of a passenger engine built by Mr. Jacob Johann, General Master Mechanic of the Wabash, St. Louis & Pacific Railway, with its specifications, and the boiler of a freight engine built for the Missouri Pacific Railway, by the Baldwin Locomotive Works, for Mr. John Hewitt, Superintendent of Motive Power, with accompanying description.

The boilers used on the steamboats on the Hudson, or North River, are of a peculiar type. With a locomotive furnace or with a cast-iron "front," with top flat or arched with two or more water legs with a set flues or large tubes with or without combustion chamber, and with a considerable length of shell; with a smoke or back connection with return tubes, or small flues overhead to a breeching or uptake. The distinguishing feature from the English practice, being the use of flat stayed surface in the furnace and length of barrel, while the English or regular marine boiler uses a cylin-

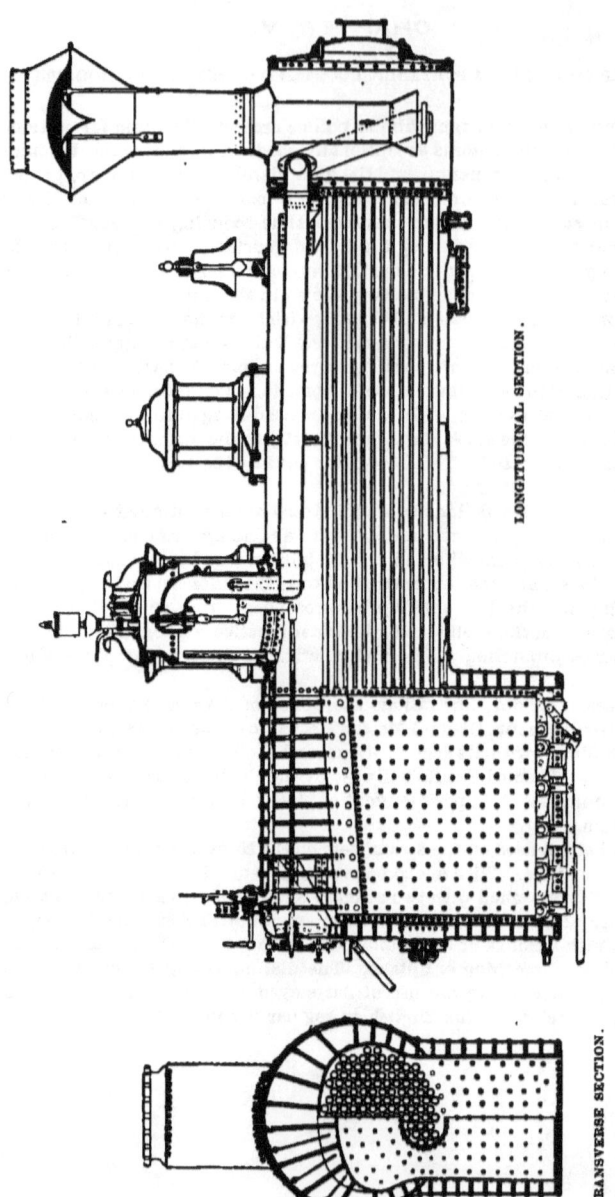

STRAIGHT TOP BOILER.

Engine No. 152, Wabash, St. Louis & Pacific Railway.

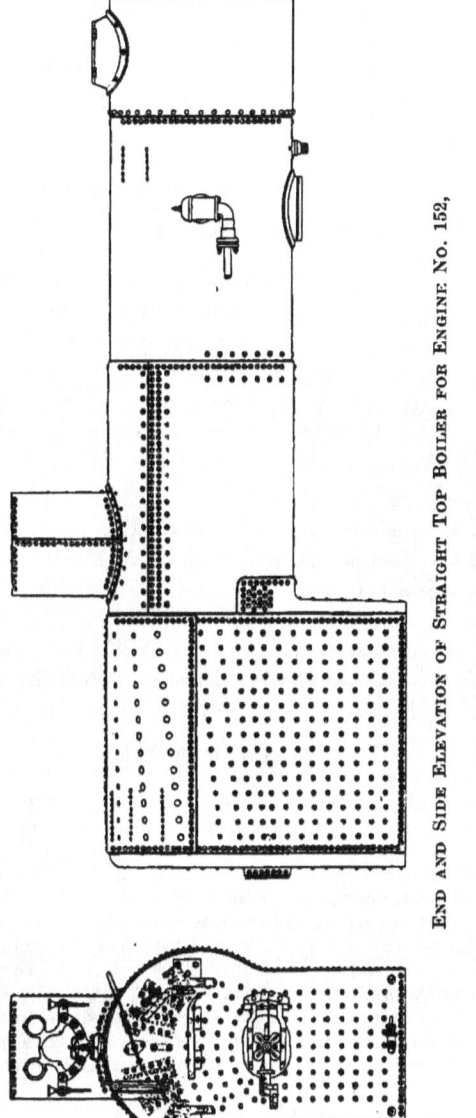

End and Side Elevation of Straight Top Boiler for Engine No. 152, Wabash, St. Louis & Pacific Railway.

drical furnace tube of large diameter and short length with back connection and with small return tubes above. According to the size of boiler there are used one, two, three, four, or six furnace tubes. The single furnace is found inconvenient for cleaning as taking in too much cold air at once, loosing steam in the operation. The size of the boiler used has been gradually increased and so also the pressure until we find on shells 16 feet in diameter pressures as high as 90 pounds per square inch carried as the working pressure.

SPECIFICATIONS FOR LOCOMOTIVE BOILER OF ENGINE NO. 152.

WABASH, ST. LOUIS & PACIFIC RAILWAY.

*General dimensions of engine.—*

| | |
|---|---|
| Cylinder | 17 × 24 inches |
| Driving wheels diameter | 5 ft. 9½ " |
| Inlet ports | 1½ × 12¼ " |
| Steam ports | 1¼ × 16 " |
| Exhaust port | 3 × 16 " |
| Width of bridges | 1¼ " |
| Tank capacity | 3,000 gals. |
| Weight on driving wheels | 51,000 lbs. |
| Weight of engine with three gauges of water | 80,000 " |

*Boiler and fire-box of steel. Tubes of wrought-iron. Fuel, soft coal.—*

| | |
|---|---|
| Diameter inside barrel of boiler | 52 inches |
| Length of fire-box | 66 " |
| Width of fire-box | 34¾ " |
| Height of fire-box at front end | 70 " |
| Height of fire-box at back end | 64½ " |
| 160 tubes 2-inch outside diameter, length | 11 ft. 6 " |
| Heating surface in fire-box less tube area | 99.5 sq. ft. |
| Heating surface tubes outside | 963.4 " |
| Total | 1062.9 " |
| Fire grate area | 15.6 " |
| Fire grate area air openings | 5.6 " |
| Ratio heating surface to grate area | 68.3 " |
| Ratio grate area to flue area | 2.47 " |
| Water capacity with two gauges or 5½-inch above crown | 1,150 gals. or 153.3 cu. ft. |
| Steam capacity under same conditions | 47.2 " |
| Total | 200.5 " |

SPECIFICATION FOR BOILER.

*Boiler.*—Straight top made throughout of best homogeneous steel plates $\frac{7}{16}$-inch thick (unless otherwise specified) and rivetted with $\frac{3}{4}$-inch rivets spaced not over 1⅞ inch between centers. All longitudinal seams double rivetted and welted, all circular seams single rivetted and welted around the bottom to above water line.

*Pressure.*—150 pounds per square inch.

*Waist.*—Inside diameter at smoke-box end 52 inches and at fire-box end 53¾ inches. Side sheet of fire-box shell ⅜-inch in thickness, and extending to the top, forming a butt joint at crown, and over these an extra crown sheet ⅜-inch in thickness is placed, extending down far enough on each side to receive all the fire-box crown stays. This extra crown sheet is rivetted to the side sheet on each side of the butt joint formed by them at the crown, and near the lower edges of this extra sheet. Dome 28 inches in diameter inside and 28 inches high above boiler, fitted with cast-iron ring and cover, so arranged with slotted flanges that bolts may be used instead of studs. Center line of dome situated 7 feet 7 inches ahead of the back face of back head of boiler, the placing of dome so far ahead being necessitated by the system of staying employed.

*Smoke-box* is $52\frac{3}{4}$ inches diameter inside and 33 inches long from center line of rivets, securing the junction of smoke-box and waist, and the front end. Length of boiler over all, 20 feet 6 inches.

*Tubes.*—Of No. 11 lap-welded charcoal iron with copper ferrules at both ends. Tubes 160 in number, 2 inches outside diameter, 11 feet 6 inches long, separated by $\frac{3}{4}$-inch bridges in fire-box flue sheet, and by $\frac{1}{4}$-inch bridges in front flue sheet to provide for a more perfect water circulation.

*Fire-box.*—Of steel arched and sloping, with round corners, box 66 inches long, and $35\frac{5}{8}$ inches wide, inside. Height at front end 70 inches, at back end, $64\frac{1}{2}$ inches inside measure, top sloped 1 inch to 1 foot. All plates thoroughly annealed after flanging. The flue sheet $\frac{1}{2}$-inch, crown sheet $\frac{3}{8}$-inch, side and fire-door sheets $\frac{5}{16}$ of an inch in thickness. Water spaces $3\frac{1}{2}$ inches all round at the mud ring, increasing to 4 inches near the crown sheet. This increase made by closing in the side and fire-door sheets for the sides and back, and by setting out the throat sheet for the front. Stay bolts are $\frac{7}{8}$ of an inch in diameter screwed in and rivetted to sheets and not over $4\frac{1}{4}$ inches thick from centre to centre. Fire door opening formed by a special oval ring rivetted to outward flanges on outside and inside sheets.

*Crown Staying.*—Crown sheet arched transversely with a radius of 35 inches, with round corners of 13 inches radius at flue sheet end and 10 inches radius at fire door sheet end. Crown stays of 1-inch Sligo iron upset at one end to admit of $\frac{1}{8}$-inch thread spaced longitudinally $4\frac{7}{8}$ inches, and transversely $5\frac{1}{2}$ inches apart on the crown sheet. The stay holes in the boiler crown are so located that the stays enter both the inside and outside crowns at the same angle longitudinally. The obliquity at which they enter the outside crown transversely, especially the lower rows, being compensated for by the double sheets forming the outside crown. Even the stay having the greatest obliquity is by this means provided with ample thread connection in the boiler crown. The fire box crown is tapped out to 1 inch thread, and the boiler crown to $1\frac{1}{8}$-inch thread, the number of threads to the inch being the same in each case, special taps and reamers being employed for this purpose. The stays, after being screwed into place are cut off within $\frac{3}{16}$ of an inch of the inside and outside crowns and rivetted over with a few well directed blows of a hammer.

*Dome.*—The dome is 28 inches inside diameter and $\frac{3}{8}$ of an inch thick, the base is flanged out to fit the shell, and the shell is flanged up to fit inside the dome, a wrought-iron welded ring 5 inches by 1 inch thick is placed on the inside of the shell and rivetted to dome with rivets $1\frac{3}{4}$ inch centres, $1\frac{3}{8}$ inches from corner of flange; a second row of rivets $2\frac{3}{4}$ inches

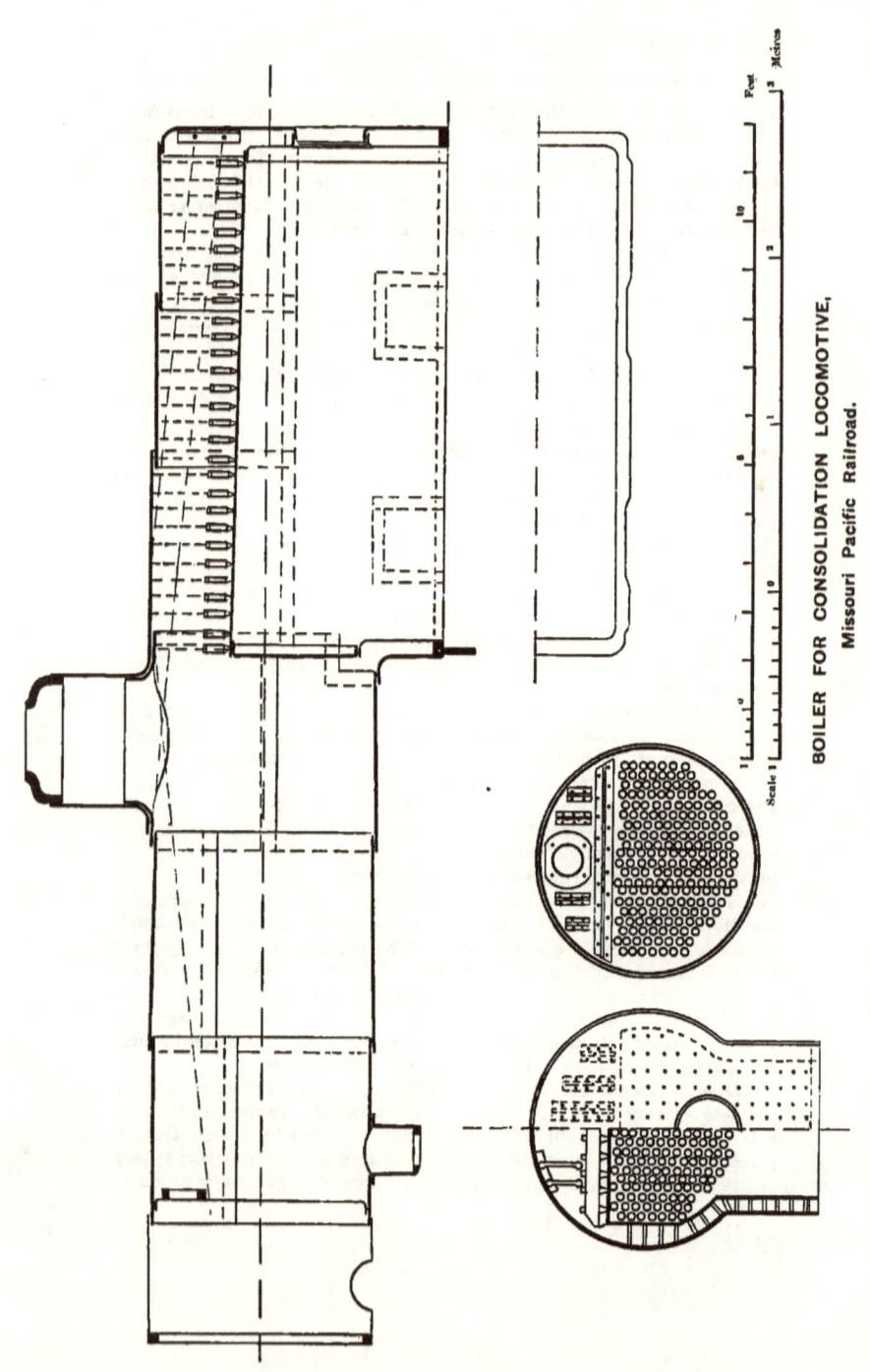

BOILER FOR CONSOLIDATION LOCOMOTIVE,
Missouri Pacific Railroad.

from flange, and 4-inch centres connects ring with shell, while a row, 4-inch centres, connects flange on shell to dome.

*Boiler Staying.*—The back head and front flue sheet are well stayed by angle gusset braces made of $\frac{7}{16}$-inch steel plates, the double angle iron connections for these braces also being made of the same material as they answer the purpose much better than if made of merchant angle iron.

*Cleaning Arrangements.*—Cylinder part of boiler is fitted with a boiler washer immediately back of front flue sheet with a man-hole just back of washer for convenience in examining boiler, cleaning plugs in corners of fire box, hand-hole plates in front leg; and blow-off cock in back end. The boiler washer consists of two horns curved to fit shell connected with nozzle provided with check valve passing through bottom of shell flange and washer secured to same. A set of nozzles point upward, and another set backwards. The boiler is always washed out under pressure and then when empty.

*Throttle.*—Balanced poppet throttle valve of cast-iron. Throttle pipe of cast-iron with flange or water shed on outside of it. Dry pipe of wrought iron 6 inches inside diameter.

*Grates.*—Of cast-iron rocking finger bar pattern, arranged to work from cab, fingers $9\frac{1}{4}$ inches long from centre of bar to end of fingers. Ash pan fitted with double dampers. All steam used except for running the engine is taken from a brass stand with one opening from boiler provided with check valve opened by an eccentric, and which closes by the pressure inside unless held open which would happen if by accident the stand was knocked off.

*Feed Water.*—Supplied by two No. 16 Rue injectors in the cab, one on each side of boiler, water enters on shell 22 inches from front head through check valves.

## DESCRIPTION OF BOILER FOR "CONSOLIDATION LOCOMOTIVE" FOR THE MISSOURI PACIFIC RAILROAD.

Shell, diameter outside smallest sheet 56; shell, length, 12 ft. $\frac{3}{4}$-in., thickness 7/16-in.; material, steel; shell of smoke box $\frac{1}{2}$-in. thick; wrought iron.

Thickness, tube sheets.................................................... $\frac{1}{2}$ inch.
" fire-box, side sheets............................................7/16 "
" fire-box, crown sheets..........................................7/16 "
" fire-box, back sheets ..........................................7/16 "
Material............................................................ Homogeneous steel.
Number of tubes....................................................198
Diameter outside of tubes..........................................2 inches.
Thickness........................................................ No. 11, B. W. G.
Fire-box stay bolts, (outside).....................................$\frac{7}{8}$ inch.
Head stay bolts, (outside).........................................1 "
Crown bar................................................6$\times\frac{5}{8}$ "
Crown stays.......................................................$\frac{7}{8}$ "

The older marine boilers were generally rectangular in form with stayed furnaces, and present type has developed from them. The only division of these boilers which we can make is based on the number of furnaces, and whether they are fired from one or both.

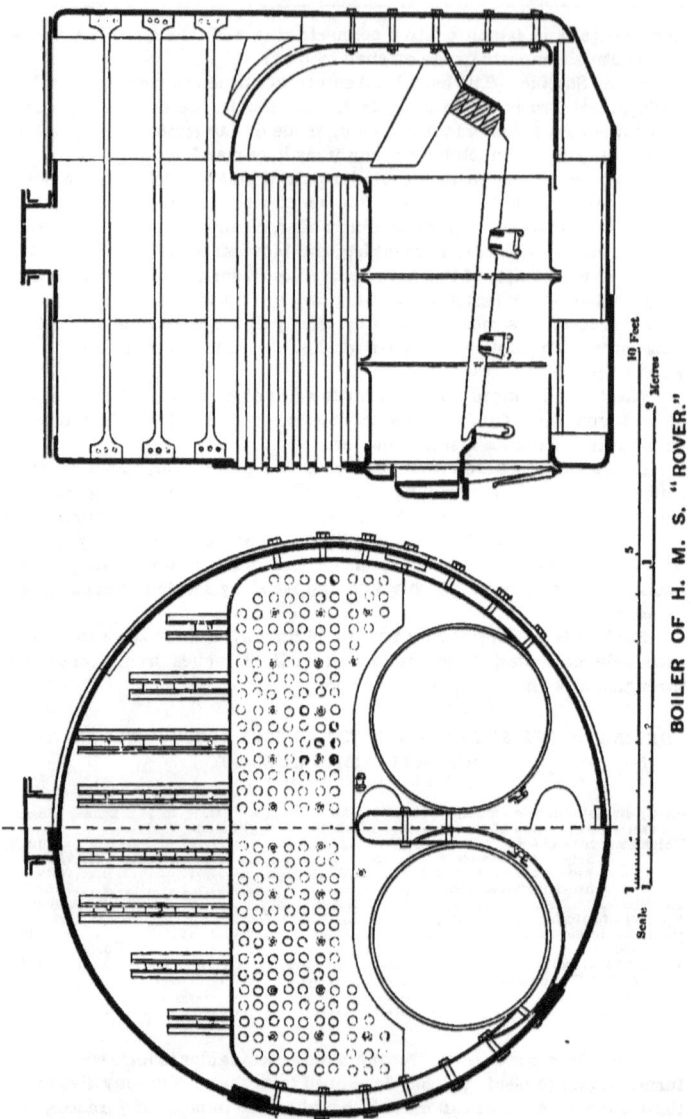

## INTERNALLY FIRED BOILERS, ETC.

Marine boilers with a single furnace are now confined to very small boilers for steam launches as it has been found by experience that it is difficult to maintain steam while the fires are being cleaned as the air introduced while cleaning lowers the steam on the entire heating surface. This type of boiler was largely used in the United States Navy, but is being replaced by those with two and three furnaces as fast as practicable.

The double furnace marine boiler which we illustrate is one of ten placed in Her Britannic Majesty's corvette "Rover." Each boiler is 11 feet 10 inches in diameter and 9 feet 6 inches long with 228 brass tubes 3 inches in diameter. The total heating surface in the ten boilers is 12,700 square feet and the total grate surface is 510 square feet. The shells are double rivetted throughout, the longitudinal seams having butt joints with butt straps inside and out. The pressure carried on trial trip was 70 pounds per square inch. The thickness of plates is not stated and the material, probably iron $\frac{7}{8}$ of an inch thick, with furnace and connections $\frac{9}{16}$ of an inch, with tube plates $\frac{3}{4}$ of an inch; end plates $\frac{3}{4}$ of an inch.

The double furnace type is usually preferred for shells 8 to 10 feet in diameter; while for larger diameters three furnaces are usually preferred.

The three furnace boiler selected as our illustration is one of the first steel marine boilers built and was selected as an exceedingly good example in every way of a very judicious design.

The diameter of shell is 13 feet 3 inches, the length 10 feet 8 inches: the shell plates are $1\frac{1}{8}$ of an inch, the end plates $\frac{9}{16}$ of an inch. The furnace and combustion chamber plates are $\frac{7}{16}$ of an inch, and the front and back tube sheets $1\frac{1}{8}$ of an inch in thickness. The screw stay bolts are $1\frac{3}{8}$ inch in diameter, and the tie-rods in the steam space are $1\frac{3}{4}$ inches in diameter with external and internal nuts and with an external washer plate rivetted to the end plate. The shell joints are all double rivetted, the longitudinal seams with double butt straps, $1\frac{1}{16}$-inch steel rivets, 4 inch centres. The lap joints are $3\frac{7}{16}$ centres. The back end plates are lapped and bent. The lap for double-rivetted seam is 5 inches. The furnace tubes are 3 feet 3 inches in external diameter and joined with double butt staps. The combustion chamber stays are 9 inch centres. The dome stands 6 ft. $10\frac{1}{2}$ in. from the shell of the boiler and is 3 ft. 8 in. inside diameter united to the shell by a short neck. The steam is taken from the dome by steam pipe, not shown in drawing. There are 244 solid drawn tubes of the same material as the shell and rivets,—that is of Landore-Siemens steel. The total heating surface is 1880 square feet and the pressure for which the boiler was designed is 65 pounds to the square inch.

The boiler was built after a long and careful series of experiments, of which we give in another place the conclusions.

Four furnace single end boilers have, we believe, been used to some extent but we have no knowledge of them. The great diameter of shell or small diameter of furnace tubes requires the use of either low pressure steam or fires of moderate thickness for which we should suggest the use of anthracite instead of soft coal.

The four furnace double-ended boiler selected for illustration, is one

## STEEL BOILER,

Built by the Wallsend Slipway Company, Newcastle-upon-Tyne.

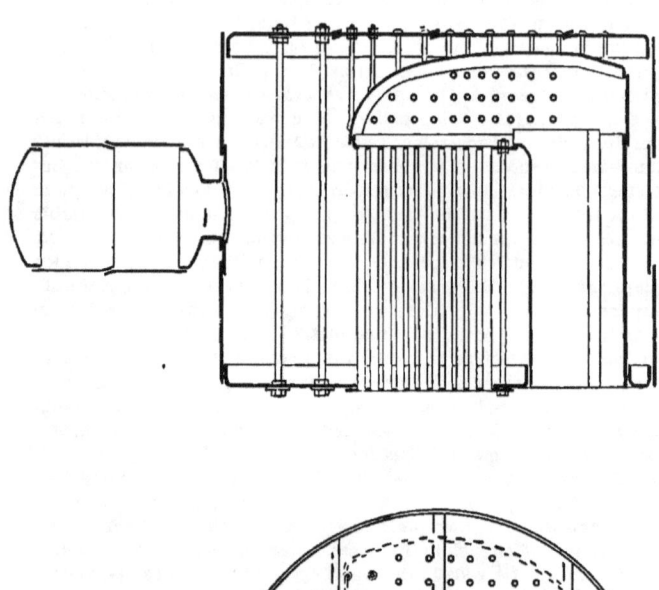

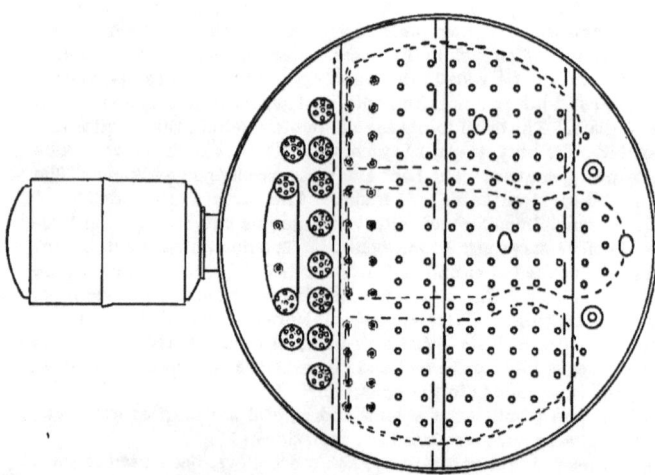

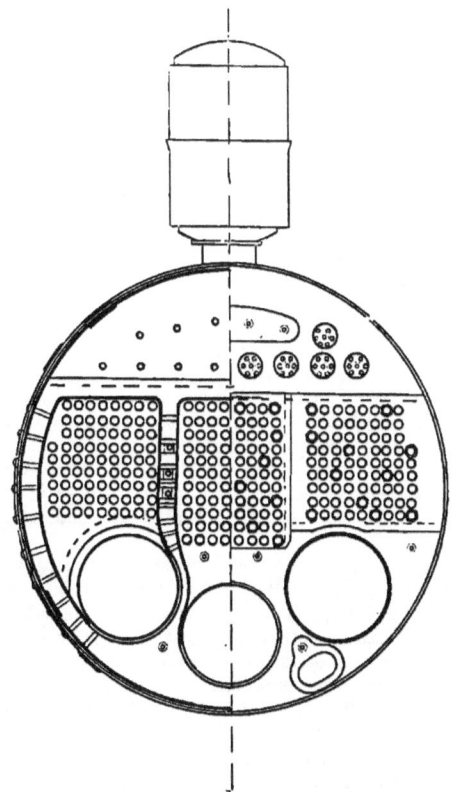

TRANSVERSE SECTION THROUGH FURNACES AND TUBES.

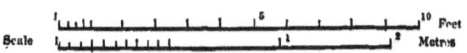

NOTE.—This scale also applies to cuts on opposite page.

of three placed in the steamship "Assyrian Monarch.' The boiler is 12 ft. 3 in. in diameter, and 18 ft. 6 in. long, with corrugated furnace tubes, 3 ft. 9 in. diameter, the latter of Don iron, the remainder is of steel, by Messrs. John Brown & Company, Sheffield. The shell plates are $1\frac{1}{16}$ of an inch, the end plates $\frac{3}{4}$ of an inch above, and $\frac{5}{8}$ of an inch below, and the tube plates are $\frac{5}{8}$ of an inch, and the corrugated tubes and combustion chambers are $\frac{1}{2}$-inch in thickness. The double-rivetted longitudinal seams have 1-inch holes and $1\frac{5}{8}$-inch rivets, $3\frac{13}{16}$ inches centres. The butt straps are 9 inches while the thickness is, for those of the central ring $1\frac{1}{8}$ of an inch inside, and $\frac{9}{16}$ of an inch outside, and for the two end rings $\frac{3}{4}$ of an inch inside, and $\frac{9}{16}$ of an inch outside. The other two rings have $\frac{9}{16}$ of an inch inside, and $1\frac{1}{16}$ of an inch outside. The ring seams have $4\frac{1}{16}$ inches thick. The rivets 1 inch in $1\frac{1}{16}$ inch holes and $3\frac{5}{8}$ inch centres double-rivetted. None of the holes are punched, but when possible, drilled in place after the sheets are bent and put together. Each ring of the shell is made of three plates. The flange plates are all annealed after flanging. The long stay rods of wrought iron, the upper row $2\frac{1}{4}$ inches and the lower row $2\frac{1}{2}$ inches in diameter. The washer plates are 9 inches in diameter and $\frac{3}{4}$ of an inch thick, rivetted to the end plates. The screw bolt stays are $1\frac{1}{2}$ inch external, $1\frac{9}{16}$ inch effective diameter of steel screwed into the plates and with nuts also at each end. There are 388 wrought iron tubes $3\frac{1}{4}$ inches external diameter, 6 ft. 7 in. long, 44 are stay tubes $2\frac{3}{4}$ inches inside diameter, being $\frac{9}{16}$ of an inch thick under the threads. These are screwed and headed over in the back tube plates while they have nuts inside and out at the other end.

The grates are 5 ft. 6 in. long, and have an area each 20.6 sq. ft. in each furnace or 82.5 sq. ft. for one boiler, and 247.5 sq. ft. in all. The tube surface in each boiler is 2140 sq. ft., and the total heating surface is 2601 sq. ft. or 7803 sq. ft. for the three boilers. The pressure allowed by the Board of Trade and Lloyd's is 80 pounds per square inch.

The machinery and vessel were built by Earle's Shipbuilding and Engineering Company, of Hull, for the Royal Exchange Shipping Company, of London, for the New York and London trade in 1881.

The last example of marine boilers, which we give, is one of three oval ones for the steamship "Mexican." The three are set athwart-ships with their center lines fore and aft. The boilers are each 12 feet 10 inches wide, 16 feet 6 inches high, and 17 feet 6 inches long, and work at 90 pounds pressure. There is one stack, and each boiler has three furnaces at each end 3 feet 4 inches diameter. The furnace tubes are of steel $\frac{7}{16}$-inch thick, and corrugated on Fox's patent. There is placed over the center boiler a steam drum 21 feet long and 5 feet diameter. There are in each boiler 440 tubes, 6 feet 9 inches between tube sheets, $3\frac{1}{4}$ inches outside diameter, No. 8 wire gauge. The stay tubes are $3\frac{1}{4}$ inches, with ends swelled to $3\frac{1}{2}$ inches. The ring sheets are $\frac{7}{8}$-inch steel, and are in length in order, 3 feet $2\frac{3}{4}$ inches, 3 feet $10\frac{5}{8}$ inches, 4 feet $7\frac{1}{4}$ inches, 3 feet $10\frac{5}{8}$ inches, and 3 feet $2\frac{3}{4}$ inches. The top end plates are $1\frac{1}{8}$-inch, with a re-enforce of $\frac{1}{2}$-inch, shown by dotted lines. The end tube plates are $\frac{3}{4}$-inch thick,

# INTERNALLY FIRED BOILERS, ETC.

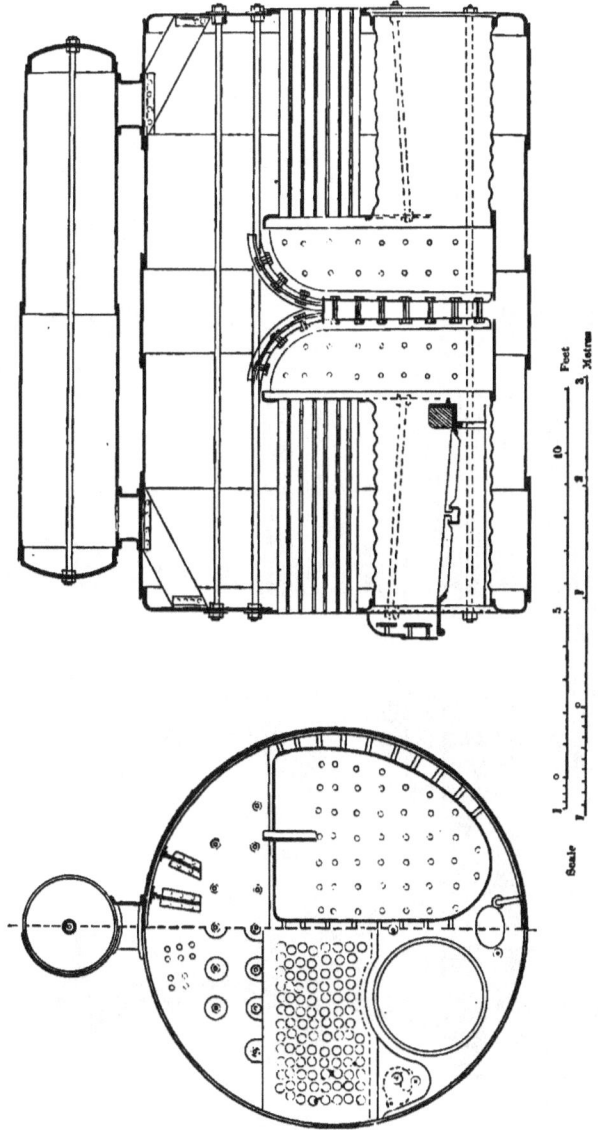

BOILER FOR S. S. "ASSYRIAN MONARCH."

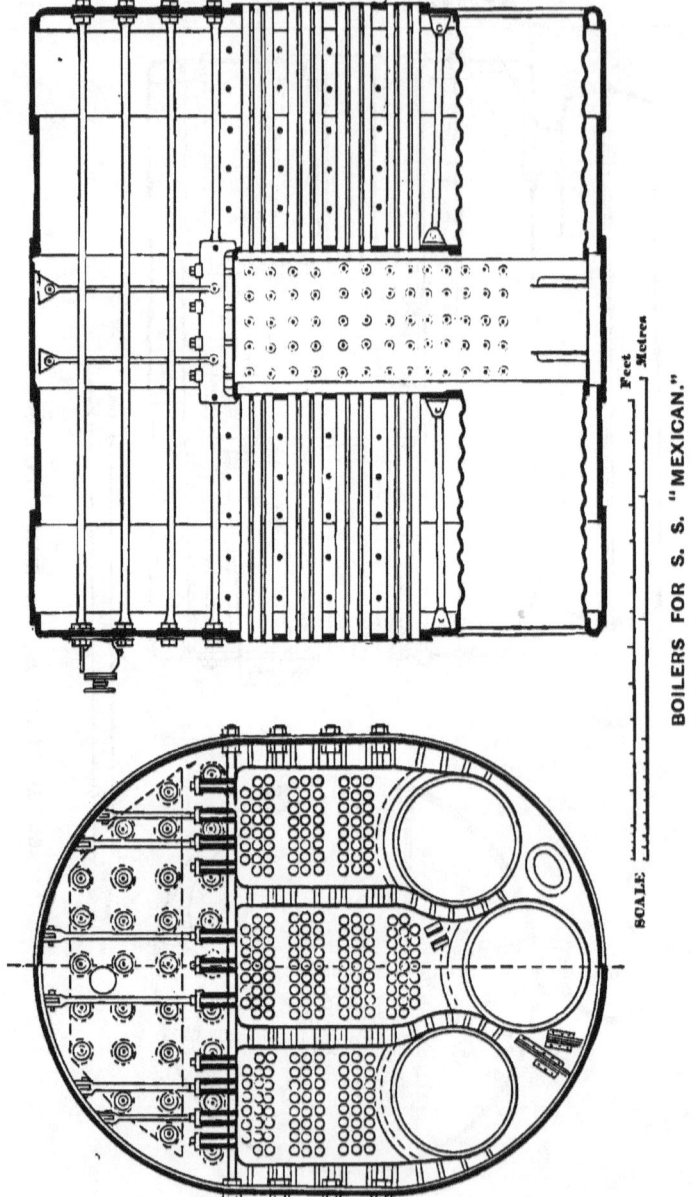

BOILERS FOR S. S. "MEXICAN."

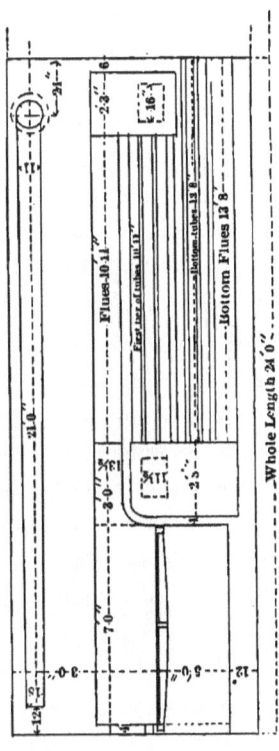
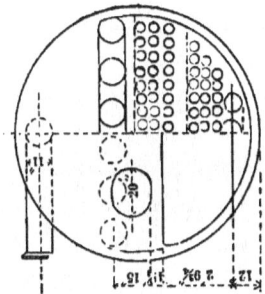
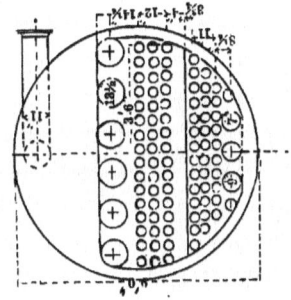

**DROP FLUE AND TUBULAR BOILER FOR THE HOBOKEN FERRY COMPANY'S BOAT, "C. LIVINGSTON."**

CYLINDER 42″ × 10″.

NEW YORK IRON WORKS, 1876.

First tier of tubes, 48—6″ diameter × 10′ 11″ long.
Bottom tier of tubes, 36—4½″ diameter × 13′ 8″ long.

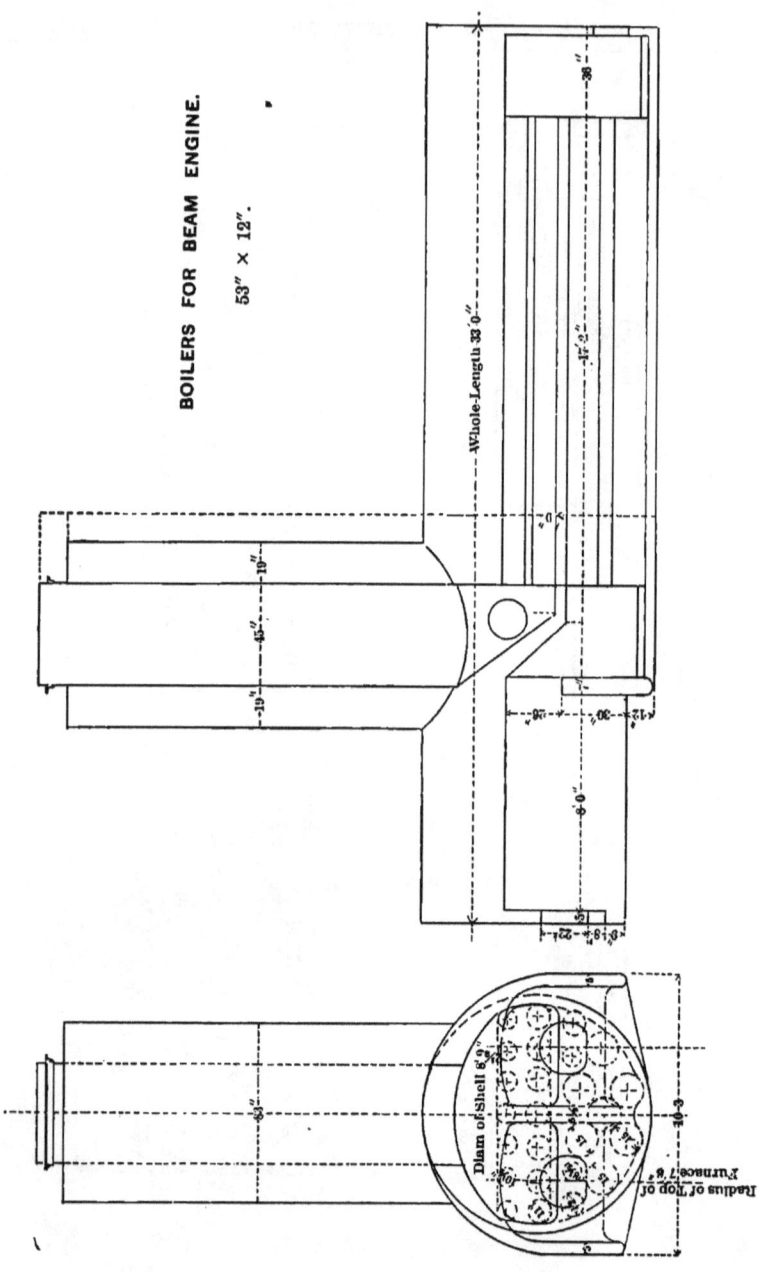

and the bottom end plates are ⅝-inch. The tube plates in the combustion chamber are ⅝-inch thick. There are thirty stay-bolts passing from head to head of boiler, with 2 7/16-inch body, and 2¾-inch screwed ends, with nuts and washers inside and out. There are forty transverse stays holding the flat sides, 2½ inches in body and 2¾-inch screwed ends, with nuts and washers inside and out. The tubes are spaced so as to make room for these last bolts. The stay tubes are screwed in the tube plates, but do not have nuts on them. The combustion chambers are stayed by bolts 1⅜-inch at bottom of threads for the center row, and 1¼-inch for the other four rows. The side sheets of combustion chambers are ½-inch thick. The top sheets are held by bolts from crown bars 10 inches by ¾-inch double, of which five sets are slung by two 1¼-inch eye-bars with fork-eyes from the sheet. The center of the drum is 6 feet 3 inches, and is of ½-inch metal.

In the three boilers there are 10,000 square feet of heating surface and 3,100 cubic feet of steam room, including the drum, and 404.5 square feet of grate.

A man-head at each end gives admission under the furnaces, and it is to be supposed there is to be admission above them. The water spaces between the tubes are 12 inches in the clear, which admits of a man passing from the top of the furnace up to the upper portion of the shell. The boilers are the best examples of accessibility that we have seen. They were built by the Southwick Engine Works, Sunderland, in 1882.

For large boilers we find this type quite a favorite one, and we mention the Cunard steamship "Servia," which carries six double ended boilers, and one single ended, with three furnaces,—thirty-nine in all. The furnace tubes are 4 feet 2 inches by 6 feet 9 inches long, with a grate surface of 1,050 square feet, and a total heating surface of 27,000 square feet. The boilers are 14 feet 10 inches wide, 18 feet high, and 18 feet 3 inches long, of Siemens steel. The furnace tubes are corrugated and are also of steel. The working pressure is 90 pounds.

Most steam fire engine boilers are of the upright tubular class. The London builders, Messrs. Shand & Mason, use a rectangular chamber traversed by nearly horizontal tubes above a circular fire-box and of course a circular shell. The Silsby Company, of New York, use a vertical tubular with hanging tubes in the fire-box, with internal circulating tubes therein; but the most successful in practice has been the "Latta," built by the Ahrens Manufacturing Company, of Cincinnati: this boiler has been used for over twenty-eight years, and although the use of a fire engine is not continuous, yet eighteen of these boilers were in continuous service for forty-nine hours at a fire in Cincinnati, on July 7 and 8, 1882, without any trouble. Their service in St. Louis, where the water is very muddy, has been very successful, and in 1881, there were twenty-one of them in use.

The peculiar features of this boiler are clearly shown in the accompanying illustrations. The boiler consists of a double shell within which is the furnace, and between which is the steam and water space. A nest of wrought iron pipes is placed in the upper part of the furnace, forming four coils starting from a single pipe and fed from the water space by a

118  STEAM MAKING; OR, BOILER PRACTICE.

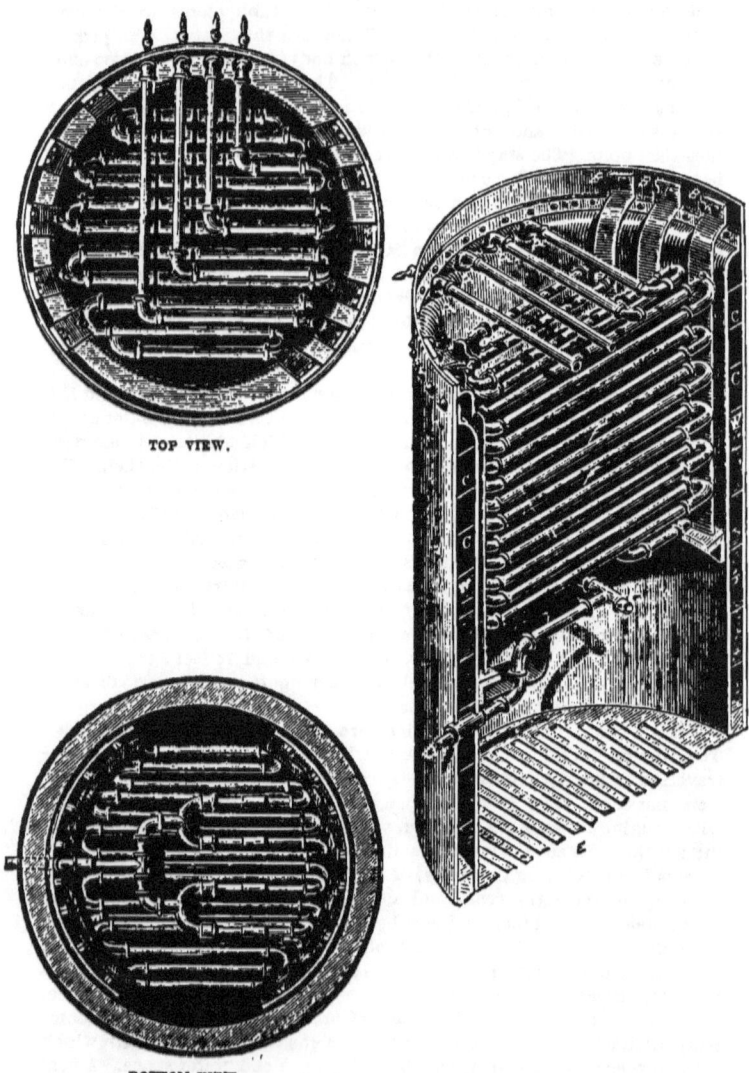

TOP VIEW.

BOTTOM VIEW.

THE "LATTA" BOILER.

separate circulating pump, discharging a mixture of steam and water into the upper part of the shell, which thus acts as a separator, the water falling through the steam to the lower part of the shell. The pipes are kept clean by the active current induced by the pump. From thirty to forty gallons of water are used at a time. A variable exhaust from the engine controls the draft, and the extent of pipe surface gives a very rapid formation of steam. The engines are guaranteed to throw water in four minutes from the lighting of the fires, with cold water.

The Herreshoff Manufacturing Company, of Bristol, R. I., have used a form of boiler for small steamboats, which differs from the "Latta" in using a spiral coil of welded pipe, a very great improvement over the jointed form shown in the "Latta," and the use of a small separator, instead of the shell, which can hardly be called an improvement for small shells. These boilers have been measurably successful with good water at sea, but we know of only three tried in Western waters. A small one on the Upper Missouri was successfully used by the United States Engineer Department for a small boat, while the other two, one at the Sabula Draw Bridge, and one in St. Louis, did not give satisfaction, soon burning out. The small boilers, the "Latta" shell and the Herreshoff coil, would probably work well for continuous service with good water, but for bad water their use will be restricted to "emergency duty."

## CHAPTER VI.

### THE DESIGN, CONSTRUCTION AND STRENGTH OF BOILERS.

The features which are most important appear to be to provide:
1. Safety; or strength.
2. Durability; or strength in the future.
3. Convenience in cleaning and inspection.
4. Capacity to do the required work. We place this last as requiring less attention than the others.

The strength of cylindrical shells is a very simple matter from the proportion. The thickness of the shell is to radius of the shell as the pressure inside the shell is to the tension around the shell. The thickness and radius being in inches, the pressure in pounds per square inch, the tension will be in pounds upon a square inch of section. The tension along the shell is one-half that around the shell.

The thickness, diameter and pressure allowed by law on steamboats in the United States is given in the accompanying extracts from the laws of the United States relative to the inspection of steamboats, taken from pamphlet form 2,100, laws governing the steamboat inspection service. Revised statutes of the United States, dated, 1882, and from the "General rules and Regulations of the Board of Supervising Inspectors of Steam Vessels," revised to 1880.

### TITLE LII. REGULATION OF STEAM VESSELS. CHAPTER I., INSPECTION.

SECTION 4399. Every vessel propelled in whole or in part by steam shall be deemed a steam vessel within the meaning of this title.

SEC. 4400. All steam vessels navigating any waters of the United States which are common highways of commerce or open to general or competitive navigation excepting public vessels of the United States, vessels of other countries and boats propelled in whole or in part by steam for navigating canals shall be subject to the provisions of this title.

SEC. 4418. The local inspectors shall also inspect the boilers of all steam vessels before the same shall be used, and once at least in every year thereafter. They shall subject all boilers to the hydrostatic pressure, and shall satisfy themselves by thorough examination that the boilers are well made, of good and suitable material; that the openings for the passage of steam and water, respectively, and all pipes and tubes exposed to heat are of proper dimensions and free from obstructions; that the spaces between and around the flues are sufficient; that the flues are circular in form; that the fire line of the furnace is at least 2 inches below the prescribed minimum low water line of the boilers; that the arrangement for delivering the feed water is such that the boilers cannot be injured thereby, and that such boilers and machinery and the appurtenances may be safely employed in the service proposed in the written application without peril to life. They shall also satisfy themselves that the safety valves are of suitable dimensions, sufficient in number and well arranged, and that the weights of the safety valves are properly adjusted so as to allow no greater pressure in the boilers than the amount prescribed in the inspection certificate; that there is a sufficient number of gauge cocks properly inserted, and to indicate the pressure of steam, suitable steam registers that will correctly record each excess of steam carried above the prescribed limit, and the highest point attained; and that there are reliable low water gauges, and that the fusible metals are properly inserted so as to fuse by the heat of the furnace whenever the water in the boilers falls below its prescribed limits: and that adequate and certain provision is made for an ample supply of water to feed the boilers at all times, whether such vessel is in motion or not, so that in high pressure boilers the water shall not be less than 4 inches above the top of the flues, and that means for blowing out are provided, so as to thoroughly remove the mud and sediment from all parts of the boilers when they are under pressure of steam. In subjecting to the hydrostatic tests boilers usually designated as high pressure boilers the inspectors shall assume 110 pounds to the square inch as the maximum pressure allowable as a working power for a new boiler of 42 inches in diameter, made in the best manner of inspected iron plates one-fourth of an inch thick, and of a quality required by law and shall rate the working power of all high pressure boilers whether old or new according to their strength compared with this standard, and in all cases the test applied shall exceed the working power allowed in the ratio of 165 to 110. In subjecting to the hydrostatic tests boilers usually designated and known as low pressure boilers, the inspectors shall allow as a working power for each new boiler a pressure of only three-fourths the number of pounds to the square inch to which it has been subjected by the hydrostatic tests, and for which it has been found to be sufficient. Should the inspectors be of the opinion that any boiler by reason of its construction or material will not safely allow so high a working pressure as is above described, they may, for reasons to be stated specially in their certificate, fix the working pressure of such boiler at less than three-fourths of the test pressure. All boilers used on steam vessels and constructed of iron or steel plates, inspected under the provisions of section 4430 shall be subjected to a hydrostatic test in the ratio of 150 to the square inch to 100 pounds to the square inch of the working steam pressure allowed. No boiler or pipe nor any of the connections therewith shall be approved which is made in whole or in part of bad material, or is unsafe in its form or dangerous from defective workmanship, age, use or other cause.

By an Act of Jan. 6, 1874, the pressure allowed boats on the Mississippi River was modified as follows:

AN ACT, Relating to the Limitation of Steam Pressure of Vessels Used Exclusively for Towing and Carrying Freight on the Mississippi River and its Tributaries.

*Be it enacted by the Senate and House of Representatives of the United States in Congress asssembled:*

That the provisions of an act entitled "An Act to Provide for the Better Security of Life on Vessels Propelled in Whole or in Part by Steam," etc., approved Feb. 28, 1871, so far as they relate to the limitation of steam pressure of steamboats used exclusively for towing and carrying freight on the Mississippi river and its tributaries, are hereby so far modified as to substitute for such boats 150 pounds of steam pressure in place of 110 pounds, as provided in said act for the standard pressure upon standard boilers of 42 inches diameter and of plates one-quarter of an inch in thickness; and such boats may, on the written permit of the supervising inspector of the district in which such boats shall carry on their business, be permitted to carry steam above the standard pressure of 110 pounds, but not exceeding the standard pressure of 150 pounds to the square inch. Approved Jan. 6th, 1874.

SEC. 4419. One of the safety valves may, if in the opinion of the local inspectors it is necessary to do so, and the steam registers shall be taken wholly from the control of all persons engaged in navigating such vessel, and secured by the inspectors.

SEC. 4426. The hull and boilers of every ferry boat, canal boat, yacht or other small craft of like character propelled by steam shall be inspected under the provisions of this title. Such other provisions of law for the better security of life as may be applicable to such vessels shall by the regulations of the board of supervising inspectors, also be required to be complied with before a certificate of inspection shall be granted and no such vessel shall be navigated without a licensed engineer, and a licensed pilot.

SEC. 4427. The hull and boiler of every tug boat, towing boat and freight boat shall be inspected under the provisions of this title, and the inspector shall see that the boilers, machinery and appurtenances of such vessel are not dangerous in form or workmanship, and that the safety valves, gauge cocks, low water alarm indicators, steam gauges and fusible plugs are all attached in conformity to law; and the officers navigating such vessels shall be licensed in conformity with the provisions of this title and shall be subject to the same provisions of law as officers navigating passenger steamers.

SEC. 4428. Every boiler manufactured to be used on steam vessels and made of iron or steel plates shall be constructed of plates that have been stamped in accordance with the provisions of the title.

SEC. 4429. Every person who constructs a boiler or steam pipe connecting the boilers, to be used on steam vessels, of iron or steel plates which have not been duly stamped and inspected according to the provisions of this title, or who knowing uses any defective, bad, or faulty iron or steel in the construction of such boilers, or who drifts any rivet hole to make it come fair; or who delivers any such boiler for use knowing it to be imperfect, in its flues, flanging, riveting, bracing, or in any other of its parts, shall be fined $1,000, one-half for the use of the informer.

Nothing in this title shall be so construed as to prevent from being used, on any steamer, any boiler or steam generator which may not be constructed of rivetted iron or steel plates, when the board of supervising inspectors have satisfactory evidence that such boiler or steam generator is equal in strength and as safe from explosion as a boiler of the best quality constructed of rivetted iron or steel plates. ["Provided, however, that the Secretary of the Treasury may grant permission to use any boiler or steam generator not constructed of rivetted iron or steel plates upon the certificate of the supervising inspector of the district wherein such boiler or generator is to be used, and other satisfactory proof that the use of the same is safe and efficient, said permit to be valid until the next regular meeting of the supervising inspectors who shall act thereon."] Amendment passed Aug. 7th, 1882.

SEC. 4430. Every iron or steel plate used in the construction of steamboat boilers, and which shall be subject to a tensile strain, shall be inspected in such manner as shall be prescribed by the Board of Supervising Inspectors and approved by the Secretary of the Treasury, so as to enable the inspectors to ascertain its tensile strength, homogeneousness, toughness and ability to withstand the effect of repeated heating and cooling; and no iron or steel plate shall be used in the construction of such boilers which has not been inspected and approved under the rules.

SEC. 4431. Every plate of boiler iron or steel made for use in the construction of steamboat boilers shall be left visible when such plates are worked into boilers, with the name of the manufacturer, the place where manufactured, and the number of pounds tensile strain it will bear to the sectional square inch; and inspectors shall keep a record in their office of the stamps upon all boiler plates and boilers which they inspect.

SEC. 4432. Every person who counterfeits or causes to be counterfeited any of the marks or stamps prescribed for boiler iron or steel plates, or who designedly stamps or causes to be stamped falsely any such plates, and every person who stamps or marks, or causes to be stamped or marked any such iron or steel plates with the name or trade mark of another with the intent to mislead or deceive shall be fined $2,000, one-half to the use of the informer, and may in addition thereto at the discretion of the court, be imprisoned not exceeding two years.

SEC. 4433. The working steam pressure allowable on boilers constructed of plates inspected as required by this title, when single-rivetted shall not produce a strain to exceed one-sixth of the tensile strength of the iron or steel plates of which such boilers are constructed; but where the longitudinal laps of the cylindrical parts of such boilers are double-rivetted, and the rivet holes have been fairly drilled instead of punched, an addition of 20 per centum to the working pressure provided for single-rivetting may be allowed. *Provided*, That all other parts of such boilers shall correspond in strength to the additional allowances so made, and no split caulking shall in any case be permitted.

SEC. 4434. No boiler to which the heat is applied to the outside of the shell thereof, shall be constructed of iron or steel plates of more than $\frac{24}{100}$ of an inch in thickness, the ends or heads of the boilers only excepted; and every such boiler employed on steam vessels navigating rivers flowing into the Gulf of Mexico, or their tributaries shall have not less than 3 inches space between and around its internal flues.

SEC. 4435. The feed water shall be delivered into the boilers in such manner as to prevent it from contracting the metal or otherwise injuring the boilers. And when boilers are so arranged on a vessel that there is employed a water connecting pipe through which the water may pass from one boiler to another, there shall also be provided a similar steam connection having an area of opening into each boiler of at least a square inch for every 2 square feet of effective heating surface contained in any one of the boilers so connected, half the flue and all other fire surfaces being computed as effective. Adequate provisions shall be made on all steam vessels to prevent sparks of flames from being driven back from the fire doors into the vessel.

SEC. 4436. Every boiler shall be provided with a good, well constructed safety valve or valves of such number, dimensions and arrangements as shall be prescribed by the Board of Supervising Inspectors and shall be also provided with a sufficient number of gauge cocks and a reliable low water indicator that will give alarm when the water falls below its prescribed limits; and in addition thereto, there shall be inserted in a suitable manner in the flues, crown sheets or other parts of the boiler most exposed to the heat of the furnace when the water falls below its prescribed limits a plug of good Banca tin.

SEC. 4437. Every person who intentionally loads or obstructs or causes to be loaded or obstructed in any way or manner, the safety valve of a boiler or who employs any other means or device whereby the boiler may be subjected to a greater pressure than the amount allowed by the certificate of the inspectors or who intentionally deranges or hinders the operation of any machinery or device employed to denote the state of the water or steam in any boiler, or to give warning of approaching danger, or who intentionally permits the water to fall below the prescribed low water line of the boiler, and every person concerned therein directly or indirectly, shall be guilty of a misdemeanor and shall be fined $200 and may also be imprisoned not exceeding five years.

## EXTRACTS FROM THE RULES AND REGULATIONS OF THE BOARD OF SUPERVISING INSPECTORS OF STEAM VESSELS.

### TABLE OF PRESSURES ALLOWABLE ON BOILERS MADE SINCE FEBRUARY 28, 1872.

| Diameter of Boiler. | Thickness of Plates. | 45,000 TENSILE STRENGTH. 1-6, 7,500. | | 50,000 TENSILE STRENGTH. 1-6, 8,333.3. | | 55,000 TENSILE STRENGTH. 1-6, 9,166.6. | | 60,000 TENSILE STRENGTH. 1-6, 10,000. | | 65,000 TENSILE STRENGTH. 1-6, 10,833.3. | | 70,000 TENSILE STRENGTH. 1-6, 11,666.6. | |
|---|---|---|---|---|---|---|---|---|---|---|---|---|---|
| | | Pressure. | 20 per cent. additional. | Pressure. | 20 per cent. additional. | Pressure. | 20 per cent. additional. | Pressure. | 20 per cent. additional. | Pressure. | 20 per cent. additional. | Pressure. | 20 per cent. additional. |
| 36 Inches. | .1875 | 78.12 | 93.74 | 86.8 | 104.16 | 95.48 | 114.57 | 104.16 | 124.99 | 112.84 | 135.4 | 121.52 | 145.82 |
| | .21 | 87.5 | 105. | 97.21 | 116.65 | 106.94 | 128.3 | 116.66 | 139.99 | 126.38 | 151.65 | 136.11 | 163.33 |
| | .23 | 95.83 | 114.99 | 106.47 | 127.76 | 117.12 | 140.54 | 127.77 | 153.32 | 138.41 | 166.09 | 149.07 | 178.88 |
| | .25 | 104.16 | 124.99 | 115.74 | 138.88 | 127.31 | 152.77 | 138.88 | 166.65 | 150.46 | 180.55 | 162.03 | 193.43 |
| | .26 | 108.33 | 129.99 | 120.37 | 144.44 | 132.4 | 158.88 | 144.44 | 173.32 | 156.48 | 187.77 | 168.51 | 202.21 |
| | .29 | 120.83 | 144.99 | 134.25 | 161.11 | 147.68 | 177.21 | 161.11 | 193.33 | 174.53 | 209.43 | 187.90 | 225.48 |
| | .3125 | 130.2 | 156.24 | 144.67 | 173.6 | 159.14 | 190.96 | 173.6 | 208.32 | 188.07 | 225.68 | 202.5 | 243.04 |
| | .33 | 137.5 | 165. | 152.77 | 183.32 | 168.05 | 201.66 | 183.33 | 219.99 | 198.61 | 238.33 | 213.88 | 256.65 |
| | .35 | 145.83 | 174.99 | 162.03 | 194.43 | 178.23 | 213.87 | 194.44 | 233.32 | 210.64 | 252.76 | 226.84 | 272.20 |
| | .375 | 156.25 | 187.5 | 173.61 | 208.33 | 190.97 | 229.16 | 208.33 | 249.99 | 225.69 | 271.82 | 243.05 | 291.66 |
| 38 Inches. | .1875 | 74.01 | 88.89 | 82.23 | 98.67 | 90.46 | 108.54 | 98.68 | 118.41 | 106.9 | 128.28 | 115.13 | 138.16 |
| | .21 | 82.89 | 99.46 | 92.1 | 110.52 | 101.31 | 121.57 | 110.52 | 132.62 | 119.73 | 143.67 | 128.93 | 154.71 |
| | .23 | 90.78 | 108.93 | 100.87 | 121.04 | 110.96 | 133.15 | 121.05 | 145.26 | 131.13 | 157.35 | 141.22 | 169.46 |
| | .25 | 98.68 | 118.41 | 109.64 | 131.56 | 120.61 | 144.73 | 131.57 | 157.88 | 142.54 | 171.04 | 153.5 | 184.20 |
| | .26 | 102.63 | 123.15 | 114.03 | 136.83 | 125.43 | 150.51 | 136.84 | 164.2 | 148.24 | 177.88 | 159.64 | 191.56 |
| | .29 | 114.47 | 137.36 | 127.19 | 152.62 | 139.91 | 167.89 | 152.63 | 183.15 | 165.35 | 198.42 | 178.06 | 213.67 |
| | .3125 | 123.35 | 148.02 | 137. | 164.46 | 150.76 | 180.91 | 164.47 | 197.36 | 178.17 | 213.8 | 191.88 | 230.25 |
| | .33 | 130.26 | 156.31 | 144.73 | 173.67 | 159.2 | 191.04 | 173.68 | 208.41 | 188.15 | 225.78 | 202.62 | 243.14 |
| | .35 | 138.15 | 165.78 | 153.5 | 184.21 | 168.85 | 202.62 | 184.21 | 221.05 | 199.56 | 239.47 | 214.91 | 257.89 |
| | .375 | 148. | 177.60 | 164.73 | 197.67 | 180.81 | 217.09 | 197.36 | 236.83 | 213.81 | 256.57 | 230.26 | 276.31 |
| 40 Inches. | .1875 | 70.31 | 84.37 | 78.12 | 93.74 | 85.93 | 103.11 | 93.75 | 112.5 | 101.56 | 121.87 | 109.37 | 131.24 |
| | .21 | 78.75 | 94.50 | 87.49 | 104.98 | 96.24 | 115.48 | 105. | 126. | 113.74 | 136.48 | 122.49 | 146.98 |
| | .23 | 86.25 | 103.5 | 95.83 | 114.99 | 105.41 | 126.49 | 115. | 138. | 124.58 | 149.49 | 134.16 | 160.99 |
| | .25 | 93.75 | 112.5 | 104.16 | 124.99 | 114.58 | 137.49 | 125. | 150. | 135.41 | 162.49 | 145.83 | 174.99 |
| | .26 | 97.5 | 117. | 108.33 | 129.99 | 119.16 | 142.99 | 130. | 156. | 140.83 | 168.99 | 151.66 | 181.99 |
| | .29 | 108.75 | 130.5 | 120.83 | 144.99 | 132.91 | 159.49 | 145. | 174. | 157.08 | 188.49 | 169.16 | 202.99 |
| | .3125 | 117.18 | 140.61 | 130.2 | 156.24 | 143.22 | 171.86 | 156.25 | 187.45 | 169.27 | 203.12 | 182.29 | 218.74 |
| | .33 | 123.75 | 148.5 | 137.49 | 164.98 | 151.24 | 181.48 | 165. | 198. | 178.74 | 214.48 | 192.49 | 230.98 |
| | .35 | 131.25 | 157.5 | 145.83 | 174.99 | 160.41 | 192.49 | 175. | 210. | 189.58 | 227.49 | 204.16 | 244.99 |
| | .375 | 140.62 | 168.74 | 156.24 | 187.48 | 171.87 | 206.24 | 187.5 | 225. | 203.12 | 243.74 | 218.74 | 262.48 |
| 42 Inches. | .1875 | 66.96 | 80.35 | 74.40 | 89.28 | 81.84 | 98.20 | 89.28 | 107.13 | 96.72 | 116.06 | 104.16 | 124.99 |
| | .21 | 75. | 90. | 83.32 | 99.99 | 91.66 | 109.99 | 100. | 120. | 108.33 | 129.99 | 116.66 | 139.99 |
| | .23 | 82.14 | 98.56 | 91.23 | 109.51 | 100.39 | 120.46 | 109.52 | 131.42 | 118.65 | 142.38 | 127.77 | 153.32 |
| | .25 | 89.28 | 107.13 | 99.2 | 119.04 | 109.12 | 130.94 | 119.04 | 142.84 | 128.96 | 154.75 | 138.88 | 166.65 |
| | .26 | 92.85 | 111.42 | 103.17 | 123.8 | 113.49 | 136.18 | 123.8 | 148.56 | 134.12 | 160.94 | 144.44 | 173.32 |
| | .29 | 103.57 | 124.28 | 115.07 | 138.08 | 126.57 | 151.85 | 138.09 | 165.7 | 149.6 | 179.52 | 161.11 | 193.33 |
| | .3125 | 111.6 | 133.92 | 124. | 148.8 | 136.4 | 163.68 | 148.74 | 178.56 | 161.2 | 193.44 | 173.61 | 208.23 |
| | .33 | 117.85 | 141.42 | 130.94 | 157.12 | 114.04 | 172.84 | 157.14 | 188.56 | 170.23 | 204.27 | 183.33 | 219.99 |
| | .35 | 125. | 150. | 138.88 | 166.65 | 152.77 | 183.32 | 166.66 | 199.99 | 180.55 | 216.66 | 194.44 | 233.32 |
| | .375 | 133.92 | 160.7 | 148.8 | 178.56 | 163.68 | 196.40 | 178.57 | 214.28 | 193.45 | 232.14 | 208.33 | 249.99 |
| 44 Inches. | .1875 | 63.92 | 76.7 | 71.01 | 83.22 | 78.12 | 93.74 | 85.22 | 102.26 | 92.32 | 110.78 | 99.42 | 119.3 |
| | .21 | 71.59 | 85.9 | 79.54 | 95.44 | 87.49 | 104.98 | 95.45 | 114.54 | 103.4 | 124.08 | 111.36 | 133.63 |
| | .23 | 78.4 | 94.08 | 87.12 | 104.54 | 95.83 | 114.99 | 104.54 | 125.44 | 113.25 | 135.9 | 121.96 | 146.35 |
| | .25 | 85.22 | 102.26 | 94.69 | 113.62 | 104.16 | 124.99 | 113.63 | 136.35 | 123.1 | 147.72 | 132.56 | 159.07 |
| | .26 | 88.63 | 106.35 | 98.48 | 118.17 | 108.33 | 129.99 | 118.18 | 141.81 | 128.02 | 153.62 | 137.87 | 165.44 |
| | .29 | 98.86 | 118.63 | 109.84 | 131.80 | 120.83 | 144.99 | 131.81 | 158.17 | 142.79 | 171.33 | 153.78 | 184.53 |
| | .3125 | 106.53 | 127.83 | 118.36 | 142.03 | 130.2 | 156.24 | 142.04 | 170.44 | 153.88 | 184.65 | 165.71 | 198.85 |
| | .33 | 112.5 | 135. | 124.99 | 149.98 | 137.49 | 164.98 | 150. | 180. | 162.49 | 194.98 | 174.99 | 209.98 |
| | .35 | 119.31 | 143.17 | 132.57 | 159.08 | 145.83 | 174.99 | 159.09 | 190.9 | 172.34 | 206.8 | 185.6 | 222.72 |
| | .375 | 127.81 | 153.37 | 142.04 | 170.44 | 156.24 | 187.48 | 170.45 | 204.54 | 184.65 | 221.58 | 198.86 | 238.63 |

## TABLE OF PRESSURES, ETC., CONTINUED.

| Diameter of Boiler. | Thickness of Plates. | 45,000 TENSILE STRENGTH. 1-6, 7,500. | | 50,000 TENSILE STRENGTH. 1-6, 8,333.3. | | 55,000 TENSILE STRENGTH. 1-6, 9,166.6. | | 60,000 TENSILE STRENGTH. 1-6, 10,000. | | 65,000 TENSILE STRENGTH. 1-6, 10,833.3. | | 70,000 TENSILE STRENGTH. 1-6, 11,666.6. | |
|---|---|---|---|---|---|---|---|---|---|---|---|---|---|
| | | Pressure. | 20 per cent. additional. | Pressure. | 20 per cent. additional. | Pressure. | 20 per cent. additional. | Pressure. | 20 per cent. additional. | Pressure. | 20 per cent. additional. | Pressure. | 20 per cent. additional. |
| **46 Inches.** | .1875 | 61.14 | 73.36 | 67.93 | 81.51 | 74.72 | 89.66 | 81.51 | 97.81 | 88.31 | 105.97 | 95.1 | 114.12 |
| | .21 | 68.47 | 82.16 | 76.08 | 91.29 | 83.69 | 100.42 | 91.3 | 109.56 | 98.91 | 118.69 | 106.52 | 127.82 |
| | .23 | 75. | 90. | 83.33 | 100. | 91.66 | 109.99 | 100. | 120. | 108.33 | 129.99 | 116.66 | 139.99 |
| | .25 | 81.51 | 97.81 | 90.57 | 108.68 | 99.63 | 119.55 | 108.69 | 130.42 | 117.75 | 141.3 | 126.8 | 152.16 |
| | .26 | 84.78 | 101.73 | 94.2 | 113.04 | 103.62 | 124.34 | 113.44 | 135.64 | 122.46 | 146.05 | 131.88 | 158.25 |
| | .29 | 94.56 | 113.47 | 105.07 | 126. | 115.57 | 138.68 | 126.09 | 151.3 | 136.59 | 163.92 | 147.1 | 176.52 |
| | .3125 | 101.9 | 122.28 | 113.21 | 135.86 | 124.54 | 149.44 | 135.86 | 163.03 | 147.19 | 176.62 | 158.51 | 190.21 |
| | .33 | 107.6 | 129.12 | 119.56 | 143.47 | 131.52 | 157.82 | 143.97 | 172.16 | 155.43 | 186.51 | 167.39 | 200.86 |
| | .35 | 114.13 | 136.95 | 126.8 | 152.16 | 139.49 | 167.38 | 152.17 | 182.6 | 164.85 | 197.82 | 177.53 | 213.03 |
| | .375 | 122.28 | 146.73 | 135.86 | 163.03 | 149.45 | 179.34 | 163.04 | 195.64 | 176.62 | 211.94 | 190.21 | 228.25 |
| **48 Inches.** | .1875 | 58.59 | 70.30 | 65.1 | 78.12 | 71.61 | 85.93 | 78.12 | 93.74 | 84.63 | 101.55 | 91.13 | 109.35 |
| | .21 | 65.62 | 78.74 | 72.91 | 87.49 | 80.2 | 96.24 | 87.49 | 104.98 | 94.79 | 113.74 | 102.08 | 122.49 |
| | .23 | 71.87 | 86.24 | 79.85 | 95.82 | 87.84 | 105.4 | 95.83 | 114.99 | 103.81 | 124.57 | 111.8 | 133.16 |
| | .25 | 78.12 | 93.74 | 86.8 | 104.16 | 95.48 | 114.57 | 104.16 | 124.99 | 112.84 | 135.4 | 121.52 | 145.82 |
| | .26 | 81.25 | 97.50 | 90.27 | 108.32 | 99.3 | 119.16 | 108.33 | 129.99 | 117.36 | 140.83 | 126.38 | 151.65 |
| | .29 | 90.62 | 108.74 | 100.69 | 120.82 | 110.76 | 132.91 | 120.83 | 144.99 | 130.9 | 157.08 | 140.97 | 169.16 |
| | .3125 | 97.65 | 117.18 | 108.5 | 130.2 | 119.35 | 143.22 | 130.21 | 156.25 | 141.05 | 169.26 | 151.9 | 182.28 |
| | .33 | 103.12 | 123.74 | 114.58 | 137.49 | 126.04 | 151.24 | 137.5 | 165. | 148.95 | 178.74 | 160.41 | 192.49 |
| | .35 | 109.37 | 131.24 | 121.52 | 145.82 | 133.67 | 160.4 | 145.83 | 174.99 | 157.98 | 189.57 | 170.13 | 204.15 |
| | .375 | 117.18 | 140.61 | 130.2 | 156.24 | 143.22 | 171.86 | 156.25 | 187.50 | 169.27 | 203.12 | 182.29 | 218.74 |
| **54 Inches.** | .1875 | 52.08 | 62.49 | 57.87 | 69.44 | 63.65 | 76.38 | 69.44 | 82.44 | 75.23 | 90.27 | 81.01 | 97.21 |
| | .21 | 58.35 | 69.99 | 64.81 | 77.77 | 71.29 | 85.54 | 77.77 | 93.32 | 84.25 | 101.1 | 90.74 | 108.88 |
| | .23 | 63.88 | 76.65 | 70.98 | 85.17 | 78.08 | 93.69 | 85.18 | 102.21 | 92.28 | 110.73 | 99.38 | 119.25 |
| | .25 | 69.44 | 83.32 | 77.16 | 92.59 | 84.87 | 101.84 | 92.59 | 111.10 | 100.3 | 120.36 | 108.02 | 120.62 |
| | .26 | 72.22 | 86.66 | 80.24 | 96.28 | 88.27 | 105.92 | 96.29 | 115.54 | 104.31 | 125.17 | 112.44 | 134.8 |
| | .29 | 80.55 | 96.66 | 89.5 | 107.40 | 98.45 | 118.14 | 107.41 | 128.88 | 116.35 | 139.62 | 125.3 | 150.36 |
| | .3125 | 86.8 | 104.16 | 96.44 | 115.72 | 106.09 | 127.30 | 115.55 | 138.66 | 125.38 | 150.45 | 135.03 | 162.03 |
| | .33 | 91.66 | 109.99 | 101.84 | 122.22 | 112.03 | 134.43 | 122.22 | 146.86 | 132.4 | 158.88 | 142.59 | 171.10 |
| | .35 | 97.22 | 116.66 | 108.02 | 129.62 | 118.82 | 142.58 | 129.62 | 155.54 | 140.43 | 168.51 | 151.23 | 181.47 |
| | .375 | 104.16 | 124.99 | 115.74 | 138.88 | 127.31 | 152.77 | 138.88 | 166.65 | 150.46 | 180.55 | 162.03 | 194.43 |
| **60 Inches.** | .1875 | 46.87 | 56.24 | 52.08 | 62.49 | 57.29 | 68.74 | 62.5 | 75. | 67.7 | 81.24 | 72.91 | 87.49 |
| | .21 | 52.5 | 63. | 58.33 | 69.99 | 64.16 | 76.99 | 69.99 | 84. | 75.83 | 90.99 | 81.66 | 97.99 |
| | .23 | 57.5 | 69. | 63.88 | 76.65 | 70.27 | 84.32 | 76.66 | 91.90 | 83.05 | 99.66 | 89.44 | 107.32 |
| | .25 | 62.5 | 75. | 69.44 | 83.32 | 76.38 | 91.65 | 83.33 | 99.99 | 90.27 | 108.32 | 97.22 | 116.66 |
| | .26 | 65. | 78. | 72.22 | 86.66 | 79.44 | 95.32 | 86.66 | 103.99 | 93.88 | 112.65 | 101.11 | 121.33 |
| | .29 | 72.5 | 87. | 80.55 | 96.66 | 88.61 | 106.33 | 96.66 | 115.99 | 104.72 | 125.66 | 112.77 | 135.32 |
| | .3125 | 78.12 | 93.74 | 86.8 | 104.16 | 95.48 | 114.57 | 104.18 | 124.99 | 112.95 | 135.54 | 121.52 | 145.82 |
| | .33 | 82.5 | 99. | 91.66 | 109.99 | 100.83 | 120.99 | 109.99 | 132. | 119.16 | 142.99 | 128.33 | 153.99 |
| | .35 | 87.5 | 105. | 97.22 | 116.66 | 106.94 | 128.32 | 116.66 | 139.99 | 126.38 | 151.65 | 136.11 | 163.32 |
| | .375 | 93.75 | 112.5 | 104.16 | 124.99 | 114.58 | 137.49 | 125. | 150. | 135.41 | 162.49 | 145.83 | 175.99 |
| **66 Inches.** | .1875 | 42.61 | 51.13 | 47.34 | 56.8 | 52.07 | 62.49 | 56.81 | 68.17 | 61.55 | 73.86 | 66.28 | 79.53 |
| | .21 | 47.72 | 57.26 | 53. | 63.63 | 58.33 | 69.99 | 63.63 | 76.35 | 68.93 | 82.71 | 74.24 | 89.08 |
| | .23 | 52.27 | 62.72 | 58. | 69.69 | 63.88 | 76.65 | 69.69 | 83.62 | 75.5 | 90.6 | 81.31 | 97.57 |
| | .25 | 56.81 | 68.17 | 63.13 | 75.75 | 69.44 | 83.32 | 75.75 | 90.90 | 82.07 | 98.48 | 88.37 | 106.04 |
| | .26 | 59.09 | 70.9 | 65.65 | 78.78 | 72.22 | 86.66 | 78.78 | 94.53 | 85.35 | 102.42 | 91.91 | 110.29 |
| | .29 | 65.90 | 79.08 | 73.23 | 87.87 | 80.55 | 96.66 | 87.87 | 105.44 | 95.2 | 114.24 | 102.52 | 123.02 |
| | .3125 | 71. | 85.2 | 78.91 | 94.69 | 86.89 | 104.16 | 94.69 | 113.62 | 102.58 | 123.09 | 110.47 | 132.56 |
| | .33 | 75. | 90. | 83.33 | 99.99 | 91.66 | 109.99 | 99.99 | 120. | 108.33 | 129.99 | 116.66 | 139.99 |
| | .35 | 79.56 | 95.47 | 88.38 | 106.05 | 97.22 | 116.66 | 106. | 127.27 | 114.80 | 137.86 | 123.73 | 148.47 |
| | .375 | 85.22 | 102.26 | 94.69 | 113.62 | 104.16 | 124.99 | 113.62 | 136.34 | 123.1 | 147.72 | 132.57 | 159.08 |
| | .1875 | 39.06 | 46.87 | 43.4 | 52.08 | 47.74 | 57.28 | 52.08 | 62.49 | 56.42 | 67.70 | 60.76 | 72.91 |
| | .21 | 43.75 | 52.5 | 48.6 | 58.33 | 53.47 | 64.16 | 58.33 | 69.99 | 63.19 | 75.82 | 68.05 | 81.66 |
| | .23 | 47.91 | 57.49 | 53.24 | 63.88 | 58.56 | 70.27 | 63.88 | 76.65 | 62.21 | 83.05 | 74.53 | 89.43 |

## TABLE OF PRESSURES, ETC., CONTINUED.

| Diameter of Boiler. | Thickness of Plates. | 45,000 TENSILE STRENGTH. 1-6, 7,500. | | 50,000 TENSILE STRENGTH. 1-6, 8,333.3. | | 55,000 TENSILE STRENGTH. 1-6, 9,166.6. | | 60,000 TENSILE STRENGTH. 1-6, 10,000. | | 65,000 TENSILE STRENGTH. 1-6, 10,833.3. | | 70,000 TENSILE STRENGTH. 1-6, 11,666.6. | |
|---|---|---|---|---|---|---|---|---|---|---|---|---|---|
| | | Pressure. | 20 per cent. additional. | Pressure. | 20 per cent. additional. | Pressure. | 20 per cent. additional. | Pressure. | 20 per cent. additional. | Pressure. | 20 per cent. additional. | Pressure. | 20 per cent. additional. |
| 72 Inches. | .25 | 52.08 | 62.49 | 57.87 | 69.44 | 63.65 | 76.38 | 69.44 | 83.32 | 75.22 | 90.26 | 81.01 | 97.21 |
| | .26 | 54.16 | 64.99 | 60.18 | 72.21 | 66.2 | 79.44 | 72.22 | 86.66 | 78.24 | 93.88 | 84.25 | 101.10 |
| | .29 | 60.41 | 72.49 | 67.12 | 80.54 | 73.84 | 88.60 | 80.55 | 96.66 | 87.20 | 104.71 | 93.98 | 112.77 |
| | .3125 | 65.10 | 78.12 | 72.33 | 86.8 | 79.57 | 95.48 | 86.8 | 104.16 | 94.03 | 112.83 | 101.27 | 121.52 |
| | .33 | 68.75 | 82.5 | 76.38 | 91.65 | 84.02 | 100.82 | 91.66 | 109.99 | 99.3 | 119.16 | 106.94 | 128.32 |
| | .35 | 72.91 | 87.49 | 81.01 | 97.21 | 89.11 | 106.93 | 97.22 | 116.66 | 105.32 | 126.38 | 113.42 | 136.1 |
| | .375 | 78.12 | 93.74 | 86.8 | 104.16 | 95.48 | 114.57 | 104.16 | 124.99 | 112.84 | 135.43 | 121.53 | 145.82 |
| 78 Inches. | .1875 | 36.05 | 43.21 | 40.06 | 48.07 | 44.07 | 52.87 | 48.07 | 57.68 | 52.08 | 62.49 | 56.08 | 67.29 |
| | .21 | 40.38 | 48.45 | 44.87 | 53.84 | 49.35 | 59.22 | 53.84 | 64.60 | 58.33 | 69.99 | 62.82 | 75.38 |
| | .23 | 44.23 | 53.07 | 49.14 | 58.96 | 54.05 | 64.86 | 58.95 | 70.76 | 63.88 | 76.65 | 68.80 | 82.56 |
| | .25 | 48.07 | 57.68 | 53.41 | 64.09 | 58.76 | 70.5 | 64.4 | 76.92 | 69.44 | 83.32 | 74.78 | 89.73 |
| | .26 | 50. | 60. | 55.55 | 66.66 | 66.11 | 73.33 | 66.66 | 79.99 | 72.22 | 86.66 | 77.77 | 93.32 |
| | .29 | 55.76 | 66.91 | 61.96 | 74.35 | 68.16 | 81.79 | 74.35 | 89.22 | 80.55 | 96.66 | 86.75 | 104.1 |
| | .3125 | 60.09 | 72.1 | 66.77 | 80.12 | 73.45 | 88.14 | 80.12 | 96.14 | 86.8 | 104.16 | 93.48 | 112.17 |
| | .33 | 63.46 | 76.15 | 70.51 | 84.61 | 77.56 | 93.07 | 84.61 | 101.53 | 91.66 | 109.99 | 98.71 | 118.45 |
| | .35 | 67.3 | 80.76 | 74.78 | 89.73 | 82.26 | 98.71 | 89.74 | 107.68 | 97.22 | 116.66 | 104.70 | 125.64 |
| | .375 | 72 11 | 86.53 | 80.12 | 96.14 | 88.14 | 105.76 | 96.15 | 115.38 | 104.16 | 124.99 | 112.17 | 134.6 |
| 84 Inches. | .1875 | 33.48 | 40.17 | 37.2 | 44.68 | 40.92 | 49.1 | 44.64 | 53.56 | 48 36 | 58.03 | 52.08 | 62.49 |
| | .21 | 37.5 | 45. | 41.66 | 49.99 | 45.83 | 54.99 | 50. | 60. | 54.16 | 64.99 | 58.33 | 69.99 |
| | .23 | 41.02 | 49.22 | 45.63 | 54.75 | 50.19 | 60.22 | 54.75 | 65.71 | 59.32 | 71.18 | 63.65 | 76.38 |
| | .25 | 44.64 | 53.56 | 49.6 | 59.52 | 54.56 | 65.47 | 59 52 | 71.42 | 64.48 | 77.37 | 69.44 | 83.32 |
| | .26 | 46.42 | 55.7 | 51.58 | 61.89 | 56.74 | 68.08 | 61.9 | 74.28 | 67.05 | 80.46 | 72.22 | 86.66 |
| | .29 | 51.78 | 62.13 | 57.53 | 69.03 | 63.29 | 75.94 | 69.04 | 82.84 | 74.8 | 89.76 | 80.55 | 96.66 |
| | .3125 | 55.8 | 66.96 | 62. | 74.4 | 68.2 | 81.84 | 74.4 | 89.28 | 80.6 | 96.72 | 86.8 | 104.16 |
| | .33 | 58.92 | 70.7 | 65.47 | 78.56 | 72.02 | 86.42 | 78.57 | 94.28 | 85.11 | 102.13 | 91.66 | 109.99 |
| | .35 | 62.5 | 75. | 69.44 | 83.32 | 76.38 | 91.65 | 83.33 | 99.99 | 90.27 | 108.32 | 97.22 | 116.66 |
| | .375 | 66.96 | 80.35 | 74.4 | 89.28 | 81.84 | 98.2 | 89.28 | 107.13 | 96.72 | 116.06 | 104.16 | 124.99 |
| 90 Inches. | .1875 | 31.25 | 37.5 | 34.72 | 41.66 | 38.19 | 45.82 | 41.66 | 49.99 | 45.13 | 54.16 | 48.68 | 58.33 |
| | .21 | 35. | 42. | 38.88 | 46.65 | 42.77 | 51.32 | 46.66 | 55.99 | 50.55 | 60.66 | 54.44 | 65.32 |
| | .23 | 38.33 | 45.99 | 42.59 | 51.10 | 46.85 | 56.22 | 51.11 | 61.33 | 55.37 | 66.44 | 59.62 | 71.54 |
| | .25 | 41.66 | 49.99 | 46.29 | 55.54 | 50.92 | 61.1 | 55.55 | 66.66 | 60.18 | 72.21 | 64.81 | 77.77 |
| | .26 | 43.33 | 51.99 | 48.14 | 57.76 | 52.96 | 63.55 | 57.77 | 69.32 | 62.59 | 75.1 | 67.4 | 80.88 |
| | .29 | 48.33 | 57.99 | 53.7 | 64.44 | 59.07 | 70.8 | 64.44 | 77.32 | 69.81 | 83.77 | 75.18 | 90.21 |
| | .3125 | 52.08 | 62.49 | 57.86 | 69.43 | 63.05 | 76.38 | 69.44 | 83.32 | 75.23 | 90.27 | 81.01 | 97.21 |
| | .33 | 55. | 66. | 61.11 | 73.33 | 67.22 | 80.66 | 73.33 | 87.99 | 79.44 | 95.32 | 85.55 | 102.66 |
| | .35 | 58.33 | 69.99 | 64.81 | 77.77 | 71.29 | 85.54 | 77.77 | 93.32 | 84 25 | 101 1 | 90.72 | 108.88 |
| | .375 | 62.5 | 75. | 69.44 | 83.32 | 76.38 | 91.65 | 83.33 | 99.99 | 90.27 | 108.32 | 97.22 | 116.66 |
| 96 Inches. | .1875 | 29.29 | 35.14 | 32.55 | 39.06 | 35.8 | 42.96 | 39.06 | 46.87 | 42.31 | 50.77 | 45.57 | 54.68 |
| | .21 | 32.81 | 39.37 | 36.45 | 43.74 | 40.1 | 48.12 | 43.75 | 52.5 | 47.39 | 56.86 | 51.04 | 61.24 |
| | .23 | 35.93 | 43.12 | 39.93 | 47.91 | 43.92 | 52.7 | 47.91 | 57.49 | 51.9 | 62.28 | 55.9 | 67.08 |
| | .25 | 39.06 | 46.87 | 43.4 | 52.08 | 47.74 | 57.28 | 52.08 | 62.49 | 56.42 | 67.67 | 60.76 | 72.91 |
| | .26 | 40.62 | 48.74 | 45.14 | 54.16 | 49.65 | 59.58 | 54.16 | 64.99 | 58.78 | 70.53 | 63.19 | 75.82 |
| | .29 | 45.31 | 54.37 | 50.34 | 60.4 | 55.38 | 66.45 | 60.41 | 72.49 | 65.45 | 78.54 | 70.48 | 84.57 |
| | .3125 | 48.82 | 58.58 | 54.25 | 65.1 | 59.67 | 71.6 | 65.1 | 78.12 | 70.52 | 84.62 | 75.95 | 91.14 |
| | .33 | 51.56 | 61.87 | 57.29 | 68.74 | 63.02 | 75.62 | 68.75 | 82.5 | 74.47 | 89.36 | 80.2 | 96.24 |
| | .35 | 54.68 | 65.61 | 60.76 | 72.91 | 66.83 | 80.19 | 72.91 | 87.49 | 78.99 | 94.78 | 85.06 | 102.07 |
| | .375 | 58.58 | 70.29 | 65.1 | 78.12 | 71.61 | 85.93 | 78.12 | 93.74 | 84.63 | 101.55 | 91.14 | 109.6 |

## TABLE OF PRESSURES

Allowed under the Provisions of the Special Act of Congress relating to the Limitation of Steam Pressure of Vessels used Exclusively for Towing and Carrying Freight on the Mississippi River and its Tributaries. Approved January 6, 1874.

| Diameter | Thickness of iron (Inches) — Pressure (Lbs) | | | | | | | | | | | | |
|---|---|---|---|---|---|---|---|---|---|---|---|---|---|
| | .19 | .20 | .21 | .22 | .23 | .24 | .25 | .26 | .27 | .28 | .29 | .30 | .31 |
| 60 inches | 79.80 | 84. | 88.20 | 92.40 | 96.60 | 100.80 | 105. | 109.20 | 113.40 | 117.60 | 121.80 | 126. | 130. |
| 58 inches | 82.55 | 86.90 | 91.24 | 95.69 | 99.93 | 104.27 | 108.62 | 113. | 117.31 | 121.65 | 126. | 130.34 | 134.68 |
| 56 inches | 85.50 | 90. | 94.50 | 99. | 103.50 | 108. | 112.50 | 117. | 121.50 | 126. | 130.50 | 135. | 139.50 |
| 54 inches | 88.66 | 93.33 | 98. | 102.66 | 107.33 | 112. | 116.66 | 121.33 | 126. | 130.66 | 134.25 | 140. | 144.66 |
| 52 inches | 92.07 | 96.92 | 101.77 | 106.81 | 111.46 | 117.07 | 121.15 | 126. | 130.84 | 135.69 | 140.63 | 145.38 | 150.23 |
| 50 inches | 95.75 | 100.80 | 105.94 | 111.08 | 115.92 | 120.95 | 126. | 131.04 | 136.08 | 141.11 | 146.16 | 151.50 | 156.24 |
| 48 inches | 99.75 | 105. | 110.25 | 115.50 | 120.75 | 126. | 131.25 | 136.50 | 141.75 | 147. | 152.25 | 157.50 | 162.75 |
| 46 inches | 104.08 | 109.66 | 115.04 | 120.52 | 126. | 131.47 | 136.95 | 142.43 | 147.91 | 152.52 | 158.87 | 164.34 | 169.82 |
| 44 inches | 108.81 | 114.54 | 120.27 | 126. | 131.72 | 137.45 | 143.18 | 148.80 | 154.63 | 160.36 | 166.04 | 171.81 | 177.54 |
| 42 inches | 114. | 120. | 126. | 132. | 138. | 144. | 150. | 156. | 162. | 168. | 174. | 180. | 186. |
| 40 inches | 119.70 | 126. | 132.30 | 138.60 | 144.90 | 151.20 | 157.50 | 163.80 | 170.10 | 176.40 | 182.70 | 189. | 195.30 |
| 38 inches | 126. | 132.63 | 139.26 | 149.50 | 152.62 | 159.15 | 165.79 | 172.42 | 179.05 | 185.68 | 192.31 | 198.94 | 205.57 |
| 36 inches | 133. | 140. | 147. | 154. | 161. | 168. | 175. | 182. | 189. | 191. | 203. | 210. | 217. |
| 34 inches | 140.82 | 148.23 | 153.64 | 163.05 | 170.47 | 177.88 | 185.29 | 192.70 | 200.11 | 207.53 | 214.94 | 222.35 | 229.76 |

The above table gives the steam-pressure allowed on boilers used on freight and towing steamers, the standard pressure being 150 pounds for a boiler 42 inches diameter, and .25 of an inch thick. To find the pressure required on other size boilers, (not given in the above table,) multiply 12,600 by the thickness and divide by the radius, or half the diameter.

THE DESIGN, CONSTRUCTION, ETC.   129

TREASURY DEPARTMENT,
Document No. 57.
Steamboat Inspection.

RULES AND REGULATIONS RELATING TO PRESSURES, BOILERS, AND THE INSPECTION OF BOILER-PLATES.

RULE 1.

*Pressure Allowable on Boilers of Various Dimensions, Built Prior to Feb. 28, 1872.*

(Pressure Equivalent to the Standard Pressure for a 42-inch Boiler, ¼-inch Iron.)

| WIRE-GAUGE. | Thickness of Iron. | 34 inches in diameter. | 36 inches in diameter. | 38 inches in diameter. | 40 inches in diameter. | 42 inches in diameter. | 44 inches in diameter. | 46 inches in diameter. |
|---|---|---|---|---|---|---|---|---|
|   | Inch. | Lbs. | Lbs. | Lbs. | Lbs. | Lbs. | Lbs. | Lbs. |
| 1 | 5—16 | 169.85 | 160.41 | 151.97 | 144.37 | 137.50 | 131.25 | 125.54 |
| 2 | 14—48 | 158.52 | 149.72 | 141.84 | 134.75 | 128.33 | 122.50 | 117.17 |
| 3 | 13—48 | 147.20 | 139.03 | 131.76 | 125.12 | 119.16 | 113.75 | 108.80 |
| 4 | 1—4 | 135.88 | 128.33 | 121.57 | 115.50 | 110.00 | 105.00 | 100.43 |
| 5 | 11—48 | 124.55 | 117.63 | 111.44 | 105.87 | 100.83 | 96.25 | 92.06 |
| 6 | 10—48 | 113.23 | 106.94 | 101.31 | 96.25 | 91.66 | 87.50 | 83.69 |
| 7 | 3—16 | 101.91 | 96.24 | 91.18 | 86.62 | 82.50 | 78.75 | 75.32 |

Boilers, however, built of steel plates prior to Feb. 28, 1872, shall be deemed to have a tensile strength of 75,000 pounds to the sectional square inch, whether stamped or not, and shall be tested under the rule prescribed for boilers inspected under the provisions of section 36 of the act relating to boilers, built after the 28th of February, 1872.

TABLE OF PRESSURES ALLOWABLE ON BOILERS MADE SINCE FEBRUARY 28, 1872.

RULE 2. In the first column to the left will be found the diameter of boilers varying by 2″ from 36″ to 48″, and by 6″ from 48″ to 96″. In the second column will be found the thickness of boiler-plates, expressed in the decimal parts of an inch, and varying—by $\frac{3}{160}$″ nearly—from $\frac{3}{16}$″ to $\frac{3}{8}$″; .1875, .25, .3125, and .375 are the decimal equivalents for $\frac{3}{16}$″, $\frac{1}{4}$″, $\frac{5}{16}$″, and $\frac{3}{8}$″. The decimals, .21, .23, .26 and .29 correspond nearly to $\frac{10}{48}$″, $\frac{11}{48}$″, $\frac{12}{48}$″, and $\frac{14}{48}$″ in the table of the pressures allowable on boilers made prior to Feb. 28, 1872. At the heads of the double columns will be found the tensile strength of the plates *per square inch of section*; also ⅙ of that amount. The pressures allowable on single-riveted boilers will be found in the first divisions of the double columns under the tensile strengths and opposite the diameters and thicknesses; and in the second divisions, the pressures allowable on boilers where all the rivet-holes have been fairly drilled instead of punched, and the longitudinal laps of their cylindrical parts double-rivetted, *in the manner prescribed by law.*

The pressure for any dimension of boilers not found in the above table, can be ascertained by the following rule, viz:

Multiply one-sixth ($\frac{1}{6}$) of the lowest tensile strength found stamped on any plate in the cylindrical shell by the thickness—expressed in inches or parts of an inch—of the thinnest plate in the same cylindrical shell, and divide by the radius or half diameter—also expressed in inches—and the sum will be the pressure allowable per square inch of surface for single-rivetting, to which add 20 per centum for double-rivetting, etc.

The hydrostatic pressure applied, under the above table and rule, must be in the proportion of 150 pounds to the square inch to 100 pounds to the square inch of the working pressure allowed.

Where flat surfaces exist, the inspector must satisfy himself that the bracing and all other parts of the boiler are of equal strength with the shell, and he must also, after applying the hydrostatic test, thoroughly examine every part of the boiler to see that no weakness or fracture has been caused thereby. Inspectors must see that the flues are of proper thickness to avoid the danger of collapse. Flues of 16 inches in diameter, made after July 1, 1877, must not be less than $\frac{5}{16}$ of an inch in thickness, and in proportion for flues of a greater or less diameter.

RULE 3. Every iron or steel plate intended for the construction of boilers to be used on steam vessels shall be stamped by the manufacturer in the following manner, viz., at the diagonal corners, at a distance of about 4 inches from the edges, and also at or near the centre of the plate, with the name of the manufacturer, the place where manufactured, and the number of pounds tensile strain it will bear to the sectional square inch.

When a sheet of boiler iron is found by the inspector with one or more stamps upon the same, the inspectors shall in every such case be governed, and rate the tensile strength of iron in accordance with the lowest stamp found upon the same.

RULE 4. The manner of inspecting and testing boiler plates intended to be used in the construction of marine boilers, by the United States inspectors, shall be as follows, viz:

The inspectors shall visit places where marine boilers are being constructed, as often as possible, for the purpose of ascertaining and making a record of the stamps upon the material, its thickness, and other qualities. To ascertain the tensile strain of the plates the inspectors shall cause a piece to be taken from each sheet to be tested, the area of which shall equal one-quarter of one square inch, on all iron $\frac{5}{16}$ of an inch thick, and under; and on all iron over $\frac{5}{16}$ of an inch thick the area shall equal the square of its thickness; and the force at which the piece can be parted in the direction of the fibre or grain, represented in pounds avoirdupois—the former multiplied by four, the latter in proportion to the ratio of its area—shall be deemed the tensile strain per square inch of the plate from which the sample was taken; and should the tensile strength ascertained by the test equal that marked on the plates from which the test pieces were taken, the said plates must be allowed to be used in the construction of marine boilers; provided always that the said plates possess the other qualities required by law, viz., homogeneousness, toughness, and ability to withstand the effect of repeated heating and cooling; but should these tests prove the marks on the said plates to be overstamped, the lots from which the test plates were taken must be rejected as failing to have the strength stamped thereon. But nothing herein shall be so construed as to prevent the manufacturers from restamping such iron at the lowest tensile strain indicated by the samples, provided such restamping is done previous to the use of the plates in the manufacture of marine boilers.

To ascertain the ductility and other lawful qualities: iron of 45,000 pounds tensile strength, and under, shall show a contraction of area of 15 per cent., and each additional 1,000 pounds tensile strength shall show 1 per cent. additional contraction of area, up to and including 55,000 T. S.

In the following table will be found the widths—*expressed in hundredths of an inch*—that will equal one-quarter of one square inch of section, of the various thicknesses of boiler plates. The signs + (plus) and — (minus) indicate that the numbers against which the signs are placed are a trifle *more* or *less*, but will not, in any instance exceed $\frac{1}{1000}$ of an inch.

The gauge to be employed by inspectors and others, to determine the thickness of boiler-plates, and the widths in the table, will be any standard American gauge furnished by the Treasury Department.

$\frac{3}{16}'' = 133 -$  .26 = 96 −  .35 = 71 −
.21 = 119 −  .29 = 86 −  $\frac{3}{8}'' = 67 +$
.23 = 109 +  $\frac{5}{16}'' = 80$  $\frac{7}{16}'' = 57 -$
$\frac{1}{4}'' = 100$  .33 = 76 +  $\frac{1}{2}'' = 50$

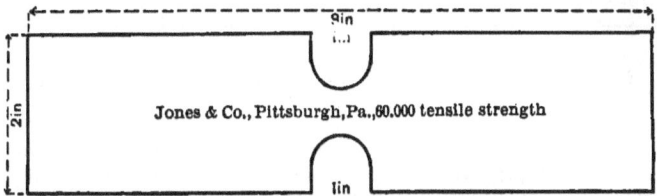

All samples intended to be tested on the Riehlé, Fairbanks, or other reliable testing machine, must be prepared in form, according to the above diagram, viz., 8 inches in length, 2 inches in width, cut out at their centres in the manner indicated.

RULE 5. The hulls and boilers of all tug, towing, and freight-boats shall be inspected in accordance with section 59 (section 4427, Revised Statutes of the act aforesaid, but steam registering gauges shall not be required on the above-named steamers.

RULE 6. The feed water shall not be admitted into any boiler, on board of any steam vessel subject to the jurisdiction of this board, at less temperature than one hundred (100) degrees Fahr. for low-pressure or condensing boilers, and one hundred and eighty (180) degrees Fahr. for high pressure or non-condensing boilers; nor shall cold water be admitted into any such boilers while the water is at a less temperature than the surrounding atmosphere.

All tests made of boiler material must be recorded upon a table of the following form:

TENSILE TESTS OF SAMPLES OF MATERIAL INTENDED TO BE EMPLOYED IN THE CONSTRUCTION OF BOILERS OF STEAM VESSELS MADE ON —— TESTING MACHINE.

| Date when tests were made. | From whom samples were obtained, and by whom tested. | Material, iron or steel. | Stamp or label on samples which must be the same as stamps on the material from which they are taken. | Thickness of samples, expressed in hundredths of an inch. | Width of samples, expressed in hundredths of an inch. | Strain at which each sample parted. | Strain per square inch of section. | Reduced thickness. | Reduced width. | Contraction of area — per cent. | REMARKS. |
|---|---|---|---|---|---|---|---|---|---|---|---|
| | | | | | | | | | | | |

RULE 7. Whenever steamers use a pressure upon their boilers exceeding 60 pounds to the square inch, they shall be inspected as high-pressure steamers and designated as such.

RULE. 8. Inspectors shall not allow the use of vertical tubular boilers, on waters flowing into the Gulf of Mexico, unless the water line of the boiler is the lawful distance above the upper end of tubes and fire line.

RULE 9. Stand pipes used on high pressure Western river steamers shall be constructed of iron, tested and stamped as required by law; but when a steamer is to be navigated in salt waters, and, in the opinion of the supervising inspector, cast iron may be employed with greater safety, he may allow stand pipes constructed of cast iron to be used on steamers of the latter class.

RULE 10. All steamers hereafter constructed, navigating the ocean, sounds, lakes, bays, and rivers, and subject to the jurisdiction of the board, shall have a clear space of not less than 16 inches on all sides of the boilers, and at the back end of all such boilers there shall be a clear space of 2 feet. Slip-joints in steam pipes shall, in their working parts, when the steamer is to be employed in navigating salt water, be made of copper or composition.

All boilers hereafter built shall have a plate or plates of sufficient size fastened on the boiler on which shall be the name of the manufacturer, the place where manufactured, and the tensile strength of the iron, and also the name of the builder of the boiler, where built, and the year.

### RELATING TO INSTRUMENTS.

RULE 33. All steam registers shall be so constructed as to be operated wholly by the pressure of steam in the boilers, to which said steam registers are attached; and their mechanism shall be enclosed in a metal case, so arranged as to preclude the possibility of its being interfered with from the outside, and be secured by a lock or such other device as the Board of Supervising Inspectors and the Secretary of the Treasury shall approve; said steam registers shall be provided with a suitable stop cock to shut off the steam from the register in case of necessity, and such stop cock shall be so placed and arranged as to be secured or locked by the same device or lock employed to secure the register from the interference of unauthorized persons. The front or face of all registers shall be of heavy glass, to enable passengers and others to observe at all times the pressure of steam. The number of times the working pressure allowed has been exceeded, and the highest point attained, may be registered either upon a continuous strip of paper under a pencil moved by the variations in the pressure of steam, or by hands or pointers moving over a dial plate or plates, arranged and graduated for that purpose.

All steam registers shall be so constructed as to record each excess of more than $2\frac{1}{2}$ pounds above the working pressure allowed for low-pressure boilers, and of more than 5 pounds above the working pressure allowed for high-pressure boilers. And each steam register, attached to either high or low-pressure boilers, shall be capable of recording at least 10 pounds above the test pressure. All steam registers shall be placed in a conspicuous position, under the direction of the inspectors, who shall also prescribe the manner in which steam registers shall be attached to boilers.

It shall be the duty of all inspectors, before issuing a certificate of inspection to any steamer, to ascertain, by actual test, that the steam register required by law is in accordance with the foregoing rules approved by the Board of Supervising Inspectors and by the Secretary of the Treasury. The makers and sellers of steam registering gauges shall furnish to the purchaser an acceptable guarantee that the same will work correctly, and according to the law and rule, one year.

RULE 34. All steamers shall have inserted in their boilers plugs of Banca tin, at least $\frac{1}{2}$-inch in diameter at the smallest end of the internal opening, in the following manner, to-wit: Cylinder boilers with flues shall have one plug inserted in one flue of each boiler; and also one plug inserted in the shell of each boiler from the inside, immediately before the fire line, and not less than 4 feet from the forward end of the boilers. All fire box boilers shall have one plug inserted in the crown of the back connection, or in the highest fire surface of the boiler. These plugs, in external diameter, must correspond in size to a 1-inch gas or steam pipe screw tap.

RULE 35. All steamers having one or two boilers shall have three suitable gauge cocks in each boiler. Those having three or more boilers in battery shall have three in each outside boiler and two in each remaining boiler in the battery; and the middle gauge cocks in all boilers shall not be less than 4 inches above the top of the flues, tubes, or crown of the fire-box.

RULE 36. Safety valves to be attached to boilers, intended for steam vessels built six months after the approval of this rule, shall have an area of not less than 1 square inch to 2 square feet of the grate surface, when the common safety valve is employed.

But when safety valves are to be used, the lift of which will give an effective area of one-half of that due the diameter of the valve, the area required shall not be less than one-half of 1 square inch to 2 square feet of the grate surface.

The valves shall be so arranged that each boiler on the steam vessel shall have one separate safety valve, unless the arrangement is such as to preclude the possibility of shutting off the communication of any boiler with the safety valve or valves employed. This arrangement shall also apply to the lock-up safety valves when they are employed.

The lock-up safety valves shall be such as are approved by the Board of Supervising Inspectors, and of such dimensions as the inspectors may deem necessary.

The term *effective area* employed in this rule has reference to the opening obtained by the lift.

The first section of this rule applies to valves constructed in material, workmanship, and principle, according to the drawings for a safety valve printed with these rules, and all common lever safety valves to be hereafter applied to the boilers of steam vessels must be so constructed.

When this construction of a safety valve is applied to the boilers of steamers navigating rough waters, the link may be connected direct with the spindle of the valve; provided always, that the fulcrum or points upon which the lever rests are made of steel, knife or sharp-edged, and hardened; in this case the short end of the lever should be attached directly to the valve-casing. In all cases the link requires but a slight movement, not exceeding $\frac{1}{8}$ of an inch.

Referring to the report on safety valve tests, conducted under the authority of the Board of Supervisors, the construction of valves which will apply to the second section of this rule is described.

All the points of bearing on lever must be in the same plane.

The distance of the fulcrum must in no case be less than the diameter of the valve-opening.

The length of the lever should not exceed the distance of the fulcrum multiplied by 10.

The width of the bearings of the fulcrum must not be less than three-fourths (¾) of one inch.

The length of the fulcrum link should not be less than four (4) inches.

The lever and fulcrum link must be made of wrought-iron or steel, and the knife-edged fulcrum points and bearings for the points must be made of steel and hardened.

The valve, valve seat, and bushings for the stem or spindle must be made of composition (gun metal) when the valve is intended to be attached to a boiler using salt water; but when the valve is to be attached to a boiler using fresh water, and generating steam of a high pressure, the parts named, with the exception of the bushings for the spindle, may be made of cast iron.

The valve must be guided by its spindle, both above and below the ground seat and above the lever, through supports either made of composition (gun metal) or bushed with it.

The spindle should fit loosely in the bearings or supports.

When the valve is intended to be applied to the boilers of steamers navigating rough waters, the fulcrum link may be connected directly with the spindle of the valve; providing always that the knife-edged fulcrum points are made of steel and hardened, and that the object sought by the link is obtained, viz., the vertical movement of the valve unobstructed by any lateral movement.

In all cases the weight must be adjusted on the lever to the pressure of steam required in each case by a correct steam gauge attached to the boiler. The weight must then be securely fastened in its position and the lever marked, for the purpose of facilitating the replacing of the weight should it be necessary to remove the same.

RULE 37. All steam gauges heretofore in use on steamers shall be admissible by the inspectors, and other steam gauges hereafter made, of equal merit, shall be allowed.

RULE 38. The appliances in use on steamers constructed prior to the 28th of February, 1872, for determining the height of water in the boilers shall be considered reliable low water gauges.

RULE 39. Every device or appliance on board of steamers which is by law, or, in the discretion of the inspectors taken from the control of all persons navigating the same, shall be secured as the supervising inspector of the district may direct, with the approval of the Secretary of the Treasury.

RULE 40. There must be means provided in all boilers using "the low water gauges" which are operated by means of a float inside the same, to prevent the float from getting into the steam pipe.

In applying the hydrostatic tests to boilers with a steam chimney, the test gauge should be applied to the "water line" of such boilers.

It shall be the duty of local inspectors to report at the end of each year to their supervising inspectors, the number of boilers inspected by them in their local districts.

All horizontal cylindrical boilers used on steamers navigating the waters flowing into the Gulf of Mexico, shall be provided with a reliable low water gauge.

RULE 41. All boilers, or sets of boilers, employed on board of vessels subject to the provisions of the act of Congress relating to steam vessels, approved February 28, 1871, for the purpose of generating steam, shall have attached to them at least one gauge that will correctly indicate the pressure of steam.

### RELATING TO DISCIPLINE.

RULE 56. It shall be the duty of an engineer, when he assumes charge of the boilers and machinery of a steamer, to forthwith thoroughly examine the same, and if he finds any part thereof in bad condition, caused by neglect or inattention on the part of his predecessor, he shall immediately report the facts to the local inspectors of the district, who shall thereupon investigate the matter, and if the former engineer has been culpably derelict of duty they shall suspend or revoke his license.

---

### INSTRUMENTS, MACHINES, AND EQUIPMENTS APPROVED FOR USE ON STEAM VESSELS.

NOTE.—Only those pertaining to boilers are here given.

### STEAM PUMPS.

Landsell's double-suction steam-siphon, presented by H. S. Landsell, New York.
Coll's single suction steam-siphon, presented by Mr. Coll, Pittsburgh.
A. Sluthouer, New Philadelphia, Ohio, fire and bilge-pump.
Coll's improved steam-siphon pump.
Sherriff's steam-siphon pump.

### DEVICES FOR REMOVING SEDIMENT FROM BOILERS.

Sediment agitator, presented by B. W. Reynolds, Evansville, Indiana.
John C. McLaughlin, Pittsburgh, Pennsylvania.
Armstrongs's vortex-skimmer.

Ordinary blow-off cocks and mud-valves.

### LOCK-UP SAFETY-VALVES.

H. G. Ashton, East Cambridge, Massachusetts.
Case & Baillie, Detroit, Michigan.
J. D. Lynde, Philadelphia, Pennsylvania.
Richardson & Co., Troy, New York.
Dry-Dock Engine Works, Detroit, Michigan.
Cockburn's safety-valve.
Ashcroft's safety-valve.
Crosby's safety valve.
Morse's safety-valve.
Hodgin's safety-valve.
A. Orme's safety-valve.

### FEED-WATER HEATERS.

Thomas Roberts, Baltimore, Maryland.
Thomas Snowden, Pittsburgh, Pennsylvania.
Doyle & Reybold, Delaware City, Delaware.
Charles G. Fisher.
H. C. Haskell, Albany, New York.
Sessler & Smith.
Heerman & Smith's water-back heater.
W. W. Martin's feed-water heater.
W. H. D. Sweet, Albany, New York.

### STEAM-GENERATORS.

Mill's auxiliary steam-generator. (Approved for use on boilers using fresh water.)

Herreshoff's patent safety coil-boilers. (Limited in construction to pipe of 4 inches in diameter.)

Farris's water-circulating grate bars, and water fronts.

### GAUGES.

Shawe's mercurial pressure.
Schaffer & Budenberg's steam-gauge.

### MISCELLANEOUS.

Torch for sailing vessels, Nathan C. Page.
Polhamus's safety-tank for fire-room floors.
Shelus's combination signal and flash light.
Gerard's electro-magnetic fire and water-detector.

## EXTRACTS FROM ENGLISH BOARD OF TRADE RULES.

In Great Britain, stationary boilers are built under the inspection of the Boiler Insurance Companies, and in some towns are subjected to municipal regulations. Locomotive boilers are usually designed by the Superintendent of Machinery, and less commonly by the manufacturer than in this country. Marine boilers are subjected to the control of the Board of Trade, if the vessels are designed to carry twelve or more passengers, but freight steamers are not subject to government inspection, but usually to Lloyds inspection. We take extracts from the rules of the Board of Trade in sufficient number to show how the strength of the shell is determined.

When cylindrical boilers are made of the best material, with all rivet holes drilled in place, and all the seams fitted with double butt straps, each of at least ⅝ the thickness of the plates they cover, and all the seams at least double rivetted with rivets having an allowance of not more than 75 per cent. over single shear, and provided that the boilers have been open to inspection during the whole period of construction, then 5 may be used as the factor of safety. The tensile strength of the iron is to be taken as equal to 47,000 pounds per square inch with the grain, and 40,000 pounds across the grain.

The boilers must be tested by hydraulic pressure to twice the working pressure, in the presence, and to the satisfaction of the Boards' surveyors. But when the above conditions are not complied with, the additions in the following scale, must be added to the factor 5, according to the circumstances of each case.

| Reference Letters. | Amount to Add. | Reason for Such Increase in Factor. |
|---|---|---|
| A | 0.15 | When all the holes are fair and in the longitudinal seams, but drilled out of place after bending. |
| B | 0.3 | When all the holes are fair and good in the longitudinal seams but drilled out of place before bending. |
| C | 0.3 | When all holes are fair and good, but are punched and not drilled after bending. |
| D | 0.5 | When the holes are fair and good, but are punched and not drilled before bending. |
| E* | 0.75 | When the holes in the longitudinal seams are not fair and good. |
| F | 0.1 | If the holes in circumferential seams are fair and good, but drilled out of place after bending. |
| G | 0.15 | If the holes in the circumferential seams are fair and good, but drilled out of place before bending. |

*Allowances E, J, W, and X, may be increased if the workmanship or material is doubtful or unsatisfactory.

EXTRACTS FROM ENGLISH BOARD OF TRADE RULES—Continued.

| Reference Letters. | Amount to Add. | Reason for Such Increase in Factor. |
|---|---|---|
| H | 0.15 | If the holes in the circumferential seams are fair and good but are punched and not drilled after bending. |
| I | 0.2 | If the holes in the circumferential seams are fair and good, but punched and not drilled before bending. |
| J* | 0.2 | If the holes in the circumferential seams are not fair and good. |
| K | 0.2 | If the longitudinal seams are double rivetted lap joints, and not double covered butts. |
| L | 0.1 | If the longitudinal seams are treble rivetted lap joints and not double covered butt straps. |
| M | 0.3 | If a single butt strap double rivetted be used in the longitudinal seams. |
| N | 0.15 | If a single butt strap treble rivetted be used in the longitudinal seams. |
| O | 1.0 | If the longitudinal seams are single rivetted, either lap single or double butt straps being used. |
| P | 0.1 | If the circumferential seams are single butt straps double rivetted. |
| Q | 0.2 | If the circumferential seams are single rivetted single butt straps. |
| R | 0.1 | If the circumferential seams are double butt straps single rivetted. |
| S | 0.1 | If the circumferential seams are double rivetted laps. |
| T | 0.2 | If the circumferential seams are single rivetted laps. |
| U | 0.25 | When the circumferential seams are lap and the straps of plates are not entirely over or under. |
| V | 0.3 | When the boiler is long and flues are used, or when it is fired from both ends, this does not affect P, R, and S. |
| W | 0.4 | If the seams are not properly crossed. |
| X* | 0.4 | If there is doubt about the metal being of the best quality. |
| Y | 1.65 | If the whole process of construction is not open to inspection. |

*Allowances E, J, W, and X, may be increased if the workmanship or material is doubtful or unsatisfactory.

It is usually stated on the authority of Sir William Fairbairn that a single-rivetted joint has 56 per cent. of the strength of the sheet, and that a double-rivetted lap has 70 per cent. of the strength of the sheet. Under these circumstances it can easily be seen that $6 \times 0.56 = 3.36$ is what is taken as the "factor of safety" by the Board of Supervising Inspectors, and that for steady pressures this is enough. Externally fired boilers built on these rules are more durable than if built with a factor of safety of 6 or 8, for the metal does not burn or the seams leak so readily. Seams leaking in the furnace cause rapid corrosion.

D. K. Clark concludes from some experiments by Brunel that for wrought-iron plates and rivets, taking the solid plate as 100, the strength of joints in plates less than $\frac{7}{8}$ of an inch thick, except in the first case, were as follows:

```
Double-rivetted butt with two straps..................................... 80
   "          lap........................................................ 72
   "          with single welt........................................... 65
Single-rivetted lap....................................................... 60
```

Concerning the strength of rivetted joints, we give some conclusions drawn by Professor A. B. W. Kennedy on steel plates and rivets from experiments made in 1881, and taken from his paper read before the Institution of Mechanical Engineers, and to be found in *Engineering*, vol. xxxi, p. 427 et. seq.

For single-rivetted lap joints the best proportions are:

    Diameter of rivet = 2.27 × thickness of plate.
    Pitch     "    = 2.22 diameter.

For double-rivetted lap joints:

    Diameter of rivet = 2.21 thickness of plate.
    Pitch     "    = 3.54 diameter.

The rivet to be $\frac{1}{16}$ of an inch smaller than the hole. The conclusion was that with steel plates and rivets the diameter and pitch for single-rivetted laps was such as would exclude their use for longitudinal seams, and that with more than $\frac{1}{2}$-inch plates the diameter of the rivet gets too large and the strength of the joint is thereby reduced. The strength of a single-rivetted lap of the proportions given above is 55 per cent., and the double-rivetted lap is 77 per cent. of the plate. The strength of the plates was 70,000 pounds tensile, and the rivets 51,000 pounds shearing stress.

We give in more detail some experiments made by Messrs. Max Eyth and David Greig at an earlier date than those of Professor Kennedy, as affording comparative data for iron and steel.

## EXPERIMENTS BY DAVID GREIG AND MAX EYTH, LEEDS, ENGLAND.

EXTRACT FROM "ENGINEERING," PAGE 527, JUNE 20, 1879.

TABLE I.—Tensile Strain of Iron and Steel Rivet Bars.

| PARTICULARS. | Mild Steel. (Brown's.) | Taylor's Iron. | Farnley Iron. | Monk Bridge B. B. Iron. | Westphalian Iron. |
|---|---|---|---|---|---|
| Diameter in inches | 0.625 | 0.625 | 0.75 | 0.75 | 0.75 |
| Area in square inches | 0.307 | 0.307 | 0.442 | 0.442 | 0.442 |
| I. Breaking strain, tons of 2,240 lbs. | 8.85 | 6.90 | 10.15 | 9.70 | 9.70 |
| II. | 8.84 | 6.75 | 10.18 | 9.90 | 9.00 |
| III. | 8.90 | 6.87 | ........ | ........ | ........ |
| Average | 8.87 | 6.84 | 10.16 | 9.80 | 9.25 |
| Breaking stress per square inch | 28.83 | 22.23 | 22.99 | 22.17 | 20.95 |
| Reduced diameter | 0.343 | 0.406 | 0.500 | 0.531 | 0.515 |
| Reduced area | 0.092 | 0.129 | 0.196 | 0.222 | 0.208 |
| Reduced to original per cent | 30 | 42 | 44 | 50 | 47 |
| A. Elongation per inch | 0.343 | 0.416 | 0.440 | 0.343 | 0.437 |
| B. " " | 0.206 | 0.279 | 0.278 | 0.225 | 0.250 |
| C. " " | 0.171 | 0.244 | 0.238 | 0.195 | 0.179 |

A. Is within the 2 inches in which the fracture occurred.
B. Is within the 10 inches in which fracture too place.
C. Is outside of the 2 inches in which fracture took place.
The tensile strength was 30 per cent. more for the steel than for the iron; the ductility less.

TABLE II.—Shearing Tests of Rivet, Iron and Steel.

(Diameter of bars ⅞-inch. Area sheared, 0.5136.)

| | IN TONS OF 2,240 POUNDS. | | |
|---|---|---|---|
| | Actual Shearing Strain. | Average. | Shear per Square Inch. |
| 1. Yorkshire Iron (Taylors) | 11.825 | | |
| 2. " " | 11.6 | | |
| 3. " " | 11.575 | 11.665 | 19.01 |
| 1. Steel (Brown) | 13.45 | | |
| 2. " " | 13.65 | | |
| 3. " " | 13.725 | 13.61 | 22.18 |

TABLE III.—Burrs Left. (Dimensions in Inches.)

| | IRON. | | STEEL. | |
|---|---|---|---|---|
| | Long Axis. | Short Axis. | Long Axis. | Short Axis. |
| No. 1 | 0.616 | 0.588 | 0.616 | 0.583 |
| No. 2 | 0.615 | 0.587 | 0.617 | 0.588 |
| No. 3 | 0.621 | 0.583 | 0.615 | 0.586 |
| Average | 0.617 | 0.586 | 0.616 | 0.581 |

### TABLE IV.—Shearing of Rivets.

(Rivets, 5/8-inch diameter. Holes, 11/16-inch diameter. Area sheared, 0.7424 sq. in.)

| Material. | Kind of Work. | IN TONS OF 2,240 POUNDS. | | |
|---|---|---|---|---|
| | | Shear on Piece. | Average. | Shear per square inch. |
| Yorkshire Iron | Hand | 14.95 | | |
| " | Hydraulic | 15.425 | | |
| " | Steam | 16.01 | 15.46 | 20.8 |
| Steel | Hand | 18.925 | | |
| " | Hydraulic | 19.320 | | |
| " | Steam | 20.4 | 19.485 | 26.3 |

### TABLE V.—Shearing of Steel Rivets.

(Rivets, 5/8-inch diameter. Holes, 11/16-inch diameter. Area sheared, 0.7424 sq. in.)

| KIND OF WORK. | Letter. | Actual Shear. | Average. | Tons of 2240 lbs. per square inch. |
|---|---|---|---|---|
| Steam Riveter | $a$ | 19.5 | | |
| | $b$ | 18.75 | | |
| | $c$ | 18.95 | 19.07 | 25.75 |
| Stationary Hydraulic | $a$ | 17.8 | | |
| | $b$ | 17.05 | | |
| | $c$ | 18.05 | 17.63 | 23.80 |
| Portable Hydraulic | $a$ | 16.70 | | |
| | $b$ | 16.85 | | |
| | $c$ | 17.11 | 16.88 | 22.78 |
| Power Light Blow | $a$ | 16.67 | | |
| | $b$ | 16.68 | 16.67 | 22.50 |
| Power Heavy Blow | $c$ | 17.6 | 17.6 | 23.76 |

The pressures on the heads of 5/8 rivets were:

```
                                                        Lbs.
Steam Rivetter..................................... 82,380
Hydraulic Stationary............................... 86,360
Hydraulic Portable................................. 44,018
Power Light Blow................................... 69,384
Power Heavy Blow................................... 115,640
```

### TABLE VI.—Shearing of Steel Rivets.

(Rivets, 5/8-inch diameter. Holes, 11/16-inch diameter. Area sheared, 0.7424 sq. in.)

| Number. | Pressure on Rivet Head. | Actual Shearing. |
|---|---|---|
| | Pounds. | Tons of 2240 lbs. |
| 1 | | 18.4 |
| 2 | 39,922 | 18.75 |
| 3 | 83,133 | 19.1 |
| 4 | 84,542 | 19.337 |
| 5 | 88,299 | 19.773 |
| 6 | | 19.95 |
| 7 | | 19.05 |
| Average. | | 19.05 |

TABLE VII.—Rivet Tests.

(Rivets, 5/8-inch diameter. Holes, 11/16-inch diameter. Area sheared, 0.7424 sq. in.)

| KIND OF WORK. | I. Steam. | II. Hydraulic. | III. Hydraulic. | IV. Power. | V. Power. |
|---|---|---|---|---|---|
| Pressure on head............ | 83280 | 86360 | 42018 | 69384 | 115640 |
| Breaking strain of sample.... | 42717 | 39491 | 37811 | 37341 | 39424 |
| Shearing strain of sample.... | 36885 | 36885 | 36885 | 36885 | 36885 |
| Friction of sample............ | 5832 | 2606 | 926 | 456 | 2539 |
| Friction of strain on one surface........................ | 2916 | 1303 | 463 | 228 | 1269 |

## CONCLUSIONS DRAWN FROM VERY EXTENDED EXPERIMENTS, BY DAVID GREIG AND MAX EYTH, OF LEEDS.

### ABSTRACTED FROM "ENGINEERING," P. 581, JUNE 27, 1879.

"It would be premature to take any of the conclusions which can be "drawn from the above tests as final, as in all practical questions, experi- "ence will have to supplement experiment, before any absolutely definite "results are arrived at. But a few facts may be pointed out, which seem "to be clearly indicated by the results of the tests, and which at least show "the direction in which further investigation may be usefully conducted, "and where practical improvements are specially required.

"There is no doubt, whatever, that the manufacturers of steel are now "able to produce a material as homogeneous and reliable as the best iron. "The absence of lamination makes it in this respect, even superior to iron "for a structure like a boiler, in which the plates are, as a rule, exposed to "strains in every direction.

"But this result has been obtained by reducing the hardness of steel "to a minimum, which materially reduces its increased usefulness. The "tensile shearing strength of the material supplied for these tests by some "of the most experienced makers of steel, and by them no doubt, consid- "ered the best for the purpose, has in these experiments proved to be not "more than 16 per cent. above that of the iron, and the want of hardness "(as distinct from tensile strength) has proved to be a very serious disad- "vantage in boiler work. What the trade now requires is "a return to a "harder material of increased tensile strength, without losing the homo- "geniety, which at present is obtained at the expense of hardness. It can "scarcely be doubted that the increasing experience in the manufacture of "steel, which has already overcome so many and such serious difficulties, "will in time meet this requirement.

"The well-known fact of the superiority of rivetting by machinery "over hand-rivetting, has been again demonstrated most conclusively, "while the experiments have shown that the effects of steam rivetting is "to say the least of it, not inferior to hydraulic rivetting as far as the quality of "the rivet is concerned but that the hydraulic rivetting is distinctly superior "as to its effects on the plate, which is less injured by the slow pressure of "the hydraulic ram.

"A number of curious facts referring to rivetted joints were indicated "by the trials. Steel showed in this respect a decided superiority over "iron beyond the proportion due to its greater tensile and shearing strength, "the average strength of all the steel seams broken being 60 per cent. of "the solid plates, that of iron only 54 per cent. This proportion was still "more striking in all lap joints, in which the greater stiffness of the ma- "terial prevented the injurious bending of the plates in the line of the "rivets, this being no doubt the chief cause of the great weakness of this "kind of joint.

"The experiments further show that the plates invariably lose part of "their tensile strength in the section of solid material left between the "rivets of a seam, this loss being greatest in lap joints. It is also greater "in punched than in drilled plates, (iron as well as steel) and greater in "plates rivetted together by steam, than in those rivetted by hydraulic "pressure. On the other hand, the strength of rivets against shearing "is greater than its normal figure, especially in lap joints.

"The usefulness of double rivetting appears to be mainly due to the "fact that it more effectually prevents lap jointed plates from bending "under stress. At the same time the zig-zag rivetting generally adopted, "in double rivetting, increases the tensile resistance of the material be- "tween the rivets considerably beyond its normal figure.

"Butt joints with a cover on one side of the plate, only gave no advan- "tage at all, the cover behaving simply as an intermediate plate attached "to the two main pieces by an ordinary lap joint. A marked improvement "could, no doubt, be obtained by giving the cover greater thickness, so as "prevent its bending.

"The most effective seams as to tensile strength, were of course, butt- "joints with two covers, as not only do they nearly double the shearing "strength of each rivet, but they entirely prevent the bending of the main "plates. The main fact resulting from the tests of parts of boilers and "complete boilers under hydraulic pressure was the impossibility of burst- "ing an ordinary rivet seam in this way, the compression of the rivet and "the elongation of the rivet hole, resulting invariably in leakage, which "prevented the necessary pressure from being obtained. Each rivet be- "comes its own safety valve, and the strain put on the weakest part of the "structure, never reached more than 70 per cent. of the breaking strain. "This is the point where additional hardness of the material would be "most useful, as it would prevent the opening of the rivet holes, which "now makes a boiler useless long before the breaking strain is reached.

"On the question of the durability of boilers it is probably impossible

TABLE VIII.—Rivetted Joints.

| Number. | | How Broken. | Breaking Strain in Pounds. Plates. | Breaking Strain in Pounds. Rivets. | Breaking Strain in Pounds. Average. | Breaking strain of solid plate ℔ in. in width. | Breaking strain of specimen ℔ in. in width. | Strength of seam per cent. of solid plate. | Width of torn section. | Strength of torn section ℔ inch of width. | Number of rivets sheared. | Shear per rivet. | Shear'g strength of rivet iron, ℔ rivet. | Strength of specimen % of normal strength. Plates. | Strength of specimen % of normal strength. Rivets. |
|---|---|---|---|---|---|---|---|---|---|---|---|---|---|---|---|
| | | | Lbs. | Lbs. | Lbs. | Lbs. | Lbs. | | Inch. | Lbs. | | Lbs. | | | |
| 1 | Single-riv. lap 6½" wide, 4 rivets ⅝" iron, holes drilled........ | Plate & riv. | 57400 | 65300 | 61350 | 18700 | 9438 | 50.4 | 3.75 | 15300 | 4 | 16325 | 15810 | 84.1 | 103 |
| 2 | " " " " " " punched...... | Plate ...... | 50800 | 48000 | 49400 | 18700 | 7600 | 40.6 | 3.5 | 14100 | | | | 75.5 | |
| 3 | " " " " steel " drilled........ | Rivets ..... | 81700 | 77900 | 79800 | 21700 | 12390 | 56.6 | | | 4 | 19950 | 18440 | | 108 |
| 4 | " " " " " " { anneal. punch. | " ..... | 86300 | 83100 | 84700 | 21700 | 12950 | 59.6 | | | 4 | 21175 | 18440 | | 115 |
| | { & unanneal. | | | | | | | | | | 4 | | | | |
| 5 | " " " " " " " | " ..... | 78600 | 91200 | 84900 | 21700 | 13060 | 60.2 | | | 4 | 21225 | 18440 | | 115 |
| 6 | " " " " iron hand work.......... | Plate ...... | 61400 | 51700 | 56550 | 18700 | 8700 | 46.5 | 3.75 | 15400 | | | | 82.5 | |
| 7 | " " " " " steam " .......... | Plate & riv. | 56600 | 67100 | 61830 | 18700 | 9515 | 50.9 | 3.75 | 15420 | 4 | 16775 | 15810 | 84.6 | 106 |
| 8 | " " " " " hyd. " .......... | " " | 61200 | 67500 | 64350 | 18700 | 9900 | 53.9 | 3.75 | 16300 | 4 | 16875 | 15810 | 89.6 | 107 |
| 9 | " " " " steel hand " .......... | " " | 72200 | 67400 | 69800 | 21700 | 10730 | 49.4 | | | 4 | 17450 | 18440 | | 94.9 |
| 10 | " " " " " steam " .......... | " " | 85700 | 79500 | 82600 | 21700 | 12707 | 58.5 | | | 4 | 20650 | 18440 | | 112 |
| 11 | " " " " " hyd. " .......... | " " | 87000 | 66400 | 76700 | 21700 | 11800 | 54.4 | | | 4 | 19175 | 18440 | | 107 |
| 12 | " 7½ " " ¾ iron.............................. | " " | 75100 | 86500 | 80800 | 18700 | 10773 | 57.6 | 4.25 | 17700 | 4 | 21625 | 22080 | 97.8 | 97.8 |
| 13 | " " " " " steel.............................. | Failed...... | | | | 21700 | | | | | | | | | |
| 14 | Double-riv. " " 6 " ⅝ iron flat zigzag ........... | Plate & riv. | 82000 | 94400 | 90700 | 18700 | 12093 | 64.6 | 5.138 | 20500 | 6 | 15733 | 15810 | 113 | 99.4 |
| 15 | " " " " " steel " .......... | " " | 110600 | 116000 | 113300 | 21700 | 15107 | 70.0 | | | 6 | 18883 | 18440 | | 102 |
| 16 | " " " 8 " ¾ iron sharp zigzag.......... | " " | 78300 | 87900 | 83100 | 18700 | 11086 | 59.2 | 4.25 | 19500 | | | | 104 | |
| 17 | " " " " " steel " .......... | " " | 96700 | 108200 | 102460 | 21700 | 13661 | 62.9 | 4.45 | 24700 | | | | 114 | |
| 18 | Butt single cover, single-riv, 6½ wide, 4 rivets ⅝" in each plate | Iron plate,. | 58000 | 58000 | 58000 | 18700 | 8923 | 46.6 | 3.75 | 15400 | | | | 82.3 | |
| 19 | " " " " " " " " | Steel rivets | 79000 | 78600 | 78800 | 21700 | 12123 | 55.8 | | | 4 | 19700 | 18440 | | 107 |
| 20 | " " double-riv., 7½ wide, 6 rivets " " | Iron plate.. | 77500 | 78500 | 78000 | 18700 | 10400 | 55.6 | 5.138 | 19400 | | | | 103 | |
| 21 | " " " " " " " | Steel rivets | 107000 | 114500 | 110750 | 21700 | 14766 | 68.0 | | | 6 | 18346 | 18440 | | 104 |
| 22 | " double cover, single-riv., 6½ wide, 4 rivets ⅝" " | Iron plate.. | 71500 | 81200 | 76350 | 18700 | 11746 | 62.7 | 3.75 | 20300 | | | | 108 | |
| 23 | " " " " " " " | Steel " .. | 93000 | 97300 | 95150 | 21700 | 131000 | 60.4 | 3.75 | 25400 | | | | 117 | |
| 24 | " " double-riv., 7½ wide, 6 rivets " " | Iron " .. | 85300 | 93000 | 89150 | 18700 | 11886 | 63.5 | 5.138 | 22150 | | | | 118 | |
| 25 | " " " " " " " | Steel " .. | 104000 | 117000 | 110500 | 21700 | 14740 | 68.0 | 5.138 | 27400 | | | | 126 | |

"to throw much light by experiments. Here practical experience is the "only reliable guide, and every well authenticated example is of some "value. The paper may therefore be concluded with the mention of one "such example which presented itself for careful examination during the "last few weeks. Two boilers similar in construction to those experimen- "ted upon were constructed by Messrs. John Fowler & Company, in the "Spring of 1868, one being entirely of steel and the other of iron. They "were used for the two engines of a steam ploughing tackle, and have just "returned for repairs to the manufacturers after eleven years of work, dur- "ing which they had been provided with new fire-boxes in 1874.

"During the whole time these boilers had to go through the severest "work and treatment to which boilers can be exposed, using every variety "of the worst water, travelling over the roughest roads and being exposed "to every sort of weather without external protection. Both boilers also "had to do exactly the same amount of work and to undergo the same "hardships as neither of the two engines can work without the other. The "result is most striking. The steel boiler has never given any trouble and "is now by far the best of the two. A few cases of this description should "finally settle the question as to the superiority of steel in this respect.

Note the conclusion regarding leakage under pressure: it is exactly what the Lancashire boiler experiment disproved, when the material was wrought-iron. We suppose the quantity of water which can be furnished in a given time is the most important element in this question.

With regard to all flat surfaces under pressure, the usual practice is to stay or tie the flat surface either to the shell or to some other more or less adjacent parallel or nearly parallel flat surface. Sometimes the flat sur- face is stiffened by rivetting an angle or T-iron on it, and often the ties used are attached by means of tee or angle irons to the flat surface, or to the shell. When the flat surface is exposed to the fire it is usual to screw the tie bolts directly into the sheet.

The conclusions arrived at by a Board of Engineers who made nu- merous experiments at the Washington Navy Yard are as follows, taken from report of the Bureau of Steam Engineering, United States Navy, for 1879:

FOR PLATES AND TIE BOLTS SCREWED THEREIN:

| Thickness of sheet. | Diameter of bolt. | Threads per inch. | Bolt projected. |
|---|---|---|---|
| $\frac{1}{4}''$ | $1''$ | 14 | $\frac{1}{2}''$ |
| $\frac{3}{8}''$ | $1\frac{1}{8}''$ | 14 | $\frac{1}{2}''$ |
| $\frac{1}{2}''$ | $1\frac{1}{4}''$ | 12 | $\frac{1}{2}''$ |
| $\frac{5}{8}''$ | $1\frac{3}{8}''$ | 12 | $\frac{1}{2}''$ |

FOR PLATES AND TIE BOLTS SCREWED THEREIN:

| IF RIVETTED TO CONE HEADS. | | IF NUTS ARE USED. | |
|---|---|---|---|
| Projection of head. | Diameter of base of cone. | Breadth of annular bearing surface. | Dished out to a depth of |
| $\frac{7}{16}''$ | $1\frac{5}{16}''$ | $\frac{1}{16}''$ | $\frac{1}{16}''$ |
| $\frac{1}{2}''$ | $1\frac{9}{16}''$ | $\frac{1}{8}''$ | $\frac{1}{16}''$ |
| $\frac{9}{16}''$ | $1\frac{3}{4}''$ | $\frac{3}{16}''$ | $\frac{3}{32}''$ |
| $\frac{5}{8}''$ | $1\frac{7}{8}''$ | $\frac{1}{4}''$ | $\frac{3}{32}''$ |

For the bursting pressure in pounds per square inch multiply the square of the quotient, arising by dividing the thickness of plate by the distance apart, center to center of bolt, by the constant which follows:

```
For iron plates and iron bolts.................................................. 192,000
For steel plates and iron bolts................................................. 200,000
For steel plates and low steel bolts............................................ 224,000
For iron plates and iron bolts with nuts....................................... 320,000
For copper plates and iron bolts................................................ 116,000
```

The large diameter of bolt used is probably intended to resist corrosion. In locomotive practice $\frac{7}{8}$-inch bolts are used for $\frac{5}{16}$-inch sheet with not more than 5-inch centres. It must not be forgotten that much of the straining and leaking at stay bolts is caused by the movements of the sheet by expansion, and that the longer the bolt the less action of this kind takes place on the sheet.

The proportion of eye-bars for tie stays was determined by the same Board, but the results are not in harmony with the practice of bridge builders in the use of eye-bars, and we shall therefore take the latter as safer practice.

The proportions found safe by experiment on full size eye-bars, are that the sum of the sectional areas of metal on each side of the eye should be 50 per cent. in excess of that in the body of the bar, and that there should be behind the eye in a plane through the centre of the bar, metal enough in the section to equal that in the body.

On the strength of flues against external pressure little is really known. Sir William Fairbairn made some experiments on this matter, and to him is due the rules commonly used, given hereafter. Mr. D. K. Clark gives other values, and a controversy took place a few years ago on the subject in the columns of *Engineering*, and we add a few experiments made by the manufacturers of the Fox Corrugated Flue. The boilers tested by the Lancashire Steam Users' Association, and by John Elder & Co., which we give in the next chapter, show flues stronger than shell, which is to be desired. The corrugations certainly have greatly increased the strength of such flues, but at present such corrugated flues are made only by one firm in England and by one in Germany. It has been proposed to make tubes of cone frustra with flange and ring joints, making a difference of say 4 to 6 inches for each foot in length. By this device sheets of $\frac{1}{4}$-inch thickness could be used up to 4 feet diameter for high pressures. There are, however, no data upon this matter. We add the experiments on full-sized flues and boilers in the next chapter.

Fairbairn's rule is that the collapsing pressure in pounds per square inch is equal to 806,000 times the 2.19 power of the thickness in inches divided by the product of the diameter in inches times the unstayed length in feet.

Fairbairn's rule as usually given, is the collapsing pressure in pounds per square inch is equal to 806,000 times the square of the thickness in inches divided by the product of the diameter in inches times the unstayed length in feet.

D. K. CLARK'S RULES.—For tubes up to 6 inches, collapsing pressure = thickness $\times \left( \dfrac{112,000}{\text{diameter}} - 12,000 \right)$. Over 6 inches, collapsing pressure = thickness $^2 \left( \dfrac{50,000}{\text{diameter}} - 5,000 \right)$. Some experiments made in Washington in 1874, agreed with the Fairbairn rule and did not give much increase of strength from an Adamson joint.

Clark states the average compression on the metal of the tube for collapsing is about 2 tons per square inch, and that the influence of length is uncertain.

## CHAPTER VII.

DESIGN AND CONSTRUCTION CONTINUED—PROPORTIONS OF HEATING SURFACE, ETC.—ECONOMIC EVAPORATION—EXPLOSIONS.

### EXPERIMENTS ON THE COLLAPSING OF FLUES.

*Experiments at Washington Navy Yard in* 1874.—An external shell with one end flanged inward and then longitudinally, and with a similar flange fitted inside the other end, was made 63 inches diameter, $\frac{3}{8}$-inch thick, and a flue $77\frac{1}{2}$ inches long and 54 inches inside diameter was rivetted to the flanges.

The flue was in two rings of $\frac{1}{4}''$ iron with an inside strap $7\frac{3}{4}'' \times \frac{1}{4}''$. Each ring was in two plates with inside strap $7\frac{3}{4}'' \times \frac{1}{4}''$, and all seams were double rivetted and caulked. Slightly oval. It was collapsed and then shored across at 105, 120, 148, 155, 180 pounds. Another flue of $\frac{1}{4}$ iron made as before, except with an Adamson joint with $1\frac{1}{8}$ ring bar, and a true cylinder: one ring collapsed at 133 pounds, and then the other at 130.

Experiments were made comparing the collapsing pressure of plain and corrugated flues.

FLUE TESTS FROM "ENGINEERING," P. 245, MARCH 29, 1878.

TABLE A.—Plain Flue.

(3' 1¾" outside diameter. 7' long, ⅜" thick.)

| | | |
|---|---|---|
| a | Horizontal diameter before pressure................................. | 36.65 inches. |
| | Vertical diameter before pressure................................... | 37.20 " |
| b | Horizontal diameter before pressure................................. | 36.62 " |
| | Vertical diameter before pressure................................... | 36.77 " |

| POUNDS PER SQUARE INCH. | POSITION IN TUBE. | DEFLECTION. | | SET. | |
|---|---|---|---|---|---|
| | | Horizontal. | Vertical. | Horizontal. | Vertical. |
| | | + | − | + | − |
| 25 | a | 0 | 0 | 0 | 0 |
| 50 | a | 0.02 | 0 | 0 | 0 |
| 75 | a | 0.02 | 0.02 | 0 | 0.02 |
| 100 | a | 0.02 | 0.04 | 0 | 0.02 |
| 25 | a | 0.02 | 0.04 | 0 | 0.02 |
| 150 | a<br>b | 0.03<br>0.00 | 0.04<br>0.03 | 0<br>0 | 0.02<br>0 |
| 175 | a<br>b | 0.03<br>0 | 0.04<br>0.05 | 0<br>0 | 0.04<br>0 |
| 200 | a | gave way | gave way | gave way | gave way |

By the Fairbairn formula this should have stood 350 pounds.

### TABLE B.—Corrugated Flue.

(Corrugations 6″ pitch, 1 5/16″ deep, metal 3/8″ thick.)

$a$ { Horizontal diameter before pressure.................................... 35.18 inches.
    { Vertical diameter before pressure....................................... 35.60 "
$b$ { Horizontal diameter before pressure.................................... 35.31 "
    { Vertical diameter before pressure....................................... 35.25 "

| POUNDS PER SQUARE INCH. | POSITION IN TUBE. | DEFLECTION. | | SET. | |
|---|---|---|---|---|---|
| | | Horizontal. | Vertical. | Horizontal. | Vertical. |
| 200 | $a$ | +0 | −0 | +0 | −0 |
| | $b$ | 0 | 0 | 0 | 0 |
| 250 | $a$ | 0 | 0.01 | 0 | 0 |
| | $b$ | 0.01 | 0.01 | 0 | 0 |
| 300 | $a$ | 0 | 0.02 | 0 | 0 |
| | $b$ | 0.01 | 0.03 | 0 | 0 |
| 350 | $a$ | 0 | 0.05 | 0 | 0 |
| | $b$ | 0.01 | 0.05 | 0 | 0 |
| 400 | $a$ | 0.05 | 0.10 | 0.04 | 0.02 |
| | $b$ | 0.02 | 0.10 | 0.04 | 0 |
| 450 | $a$ | gave way, | the weld | proved | imperfect |

Another tube of same dimensions failed with 1,070 pounds.

### TESTS MADE BY THE MANCHESTER STEAM USERS' ASSOCIATION UPON A LANCASHIRE BOILER.

EXTRACT FROM "ENGINEERING," P. 234, MARCH 24, 1876.

"The proposition to construct an experimental boiler to be tested by "hydraulic pressure up to the bursting point, was submitted to the Asso-"ciation by their Executive Committee through their chief engineer's re-"port, as far back as the month of June, 1874. Its desirability appeared "evident to the committee, especially with regard to the weakening effect "of openings cut for man-holes, steam necks, etc., which it was firmly be-"lieved tended more to the rupture of boiler shells, than was usually ad-"mitted.

"It was therefore arranged that a proper boiler should be constructed "of the diameter adopted in daily practice, and of the usual thickness "of plates with actual man-hole mouth-piece, etc., such as are common, so "that the ultimate test should be decisive.

"With this object then in view, they had a boiler constructed 21 feet "long, by 7 feet diameter inside the inner plate of shell, with two furnace

"tubes 2 feet 9 inches inside diameter, with flanged seams, each ring be-
"ing welded up solid so that there would be no rivets or lap-joints in the
"flue. The shell plates were $\frac{7}{16}$-inch thick, the ends welded up solid were
"$\frac{1}{2}$-inch thick, and the furnace tubes of $\frac{3}{8}$-inch plate, the material through-
"out being of best best Snedshill iron. All longitudinal joints were double
"rivetted and circumferential ones single rivetted. Rivet-holes were
"punched. The rivets were $2\frac{3}{8}$ inches centre to centre and the holes a mean
"diameter of $1\frac{3}{16}$-inches."

In the first experiment the boiler was complete; the man-hole mouth-piece on second-plate from back was of wrought-iron as recommended by the Association, the cast-iron pipe for blow-off elbow to attach to was also as usual, while the gusset stays and longitudinal stays were such as provided for 75 pounds working pressure. The front man-hole was placed on the inside of the front plate and furnished with usual door and cross-bar fittings. The wrought-iron neck on the top of the central sheet was the special object of the first experiment. The hole in the sheet was 17 inches in diameter, and the neck 16 inches inside by $11\frac{3}{4}$ inches high, covered with a wrought-iron plate; the neck was flanged and held on by 32 rivets. Careful records were taken of the behavior of all parts of the boiler as the pressure rose. The flat ends were carefully gauged to ascertain at what pressure and to what extent they gave away, and if any permanent set remained after the pressure was relieved. The furnace flues were carefully gauged. The length of the shell was measured by rods fixed at one end and free at the other, having pointers at several places. The circumference was measured by two encircling steel bands passed around the boiler and weighted at each end, so that by horizontal lines drawn across them, the least englargement could be rendered clearly visible.

The first rupture took place at the base of the wrought iron neck, with 250 pounds pressure; up to this point there was no movement of the ends beyond a slight permanent set which did not increase, by which fact the sufficiency of the gusset stays to secure the ends was clearly shown.

The boiler having been repaired by rivetting a thick plate over the hole whence the neck had been removed, was furnished with a cast-iron man-hole mouth-piece of the usual form. This casting was 1 inch thick, 8 inches high, and $16\frac{3}{4}$ inches inside diameter upon a 20-inch opening. The casting, which was sound, gave way with 200 lbs. per square inch pressure. For the third experiment, a dome 3 feet in diameter was used, only a small portion of the shell having been cut away, but with 235 pounds the base leaked so that no increase could be obtained. The base being stiffened by the use of rivets with heavier heads rupture of the dome flange took place with 260 pounds. No signs of trouble had been noted in the flues and ends, nor any strain.

The fifth test was made with a single rivetted joint in a longitudinal seam, which leaked with 250 pounds, and no increase of pressure could be obtained, while the double rivetted seams had all remained tight.

The sixth test was made with the ordinary oval man-hole in the shell; it was 17 inches by 13 inches, and no strengthening ring was used. Everything was new of course, and in proper order, but the head blew out of the shell with 200 pounds.

Test seven was to see if a rent could be made with the leaky seam of the single rivetted joint, with augmented pumping power; the sheet gave way along the seam and extended into both adjacent ring sheets with 275 pounds. The single rivetted seam was machine work. A double rivetted hand-made seam was then compared with the machine work, giving way with 300 pounds, while the machine work remained intact.

When the boiler was again made good, a somewhat unexpected fracture took place at the cast-iron flange connection for the blow-off, which gave way at 300 pounds, tearing the shell: after repairing this, the final test was made, when the centre seam at the bottom and middle of the boiler gave way with 310 pounds per square inch. The calculated strength of the double rivetted seams was 320 pounds. The whole of the work was done by Mr. Thomas Beeley, of Hyde Junction Iron Works, near Manchester.

FROM "ENGINEERING," PAGE 47, JULY 21, 1876.

Experiments made by Messrs. John Elder & Company, to test the ultimate strength of a boiler and super-heater removed from the S. S. "Banrigh," of the Aberdeen & London Steam Navigation Company.

The boiler was one of a pair, which with the super-heater, had been six years in service, and which were to be replaced by new ones.

## GENERAL DESCRIPTION.

The ship contained two boilers fired from both ends, the products of combustion being led over them in a flue of sheet-iron passing up through the flue of a super-heater or steam drum forming the lower part of the funnel. The boilers were made in the year 1870, and were removed after being worked hard and continuously for six years. About three years ago both boilers and super-heater underwent extensive repairs. The plates under and at the side of the bridges, the lower screwed stays, the first row of longitudinal stays over the tubes, and the tubes being removed. The shell at bottom and on each side of the water-line, showing pitting, was lined with iron plates bolted on inside. The flue of the super-heater was then entirely renewed. The dimensions of the boilers were as follows:

### BOILER.

| | |
|---|---|
| Mean diameter of shell | 11' 4½" |
| Length of shell | 12' 6" |
| Diameter of furnaces (6 in number) | 2' 10" |
| Diameter of tubes outside (404) | 2⅜" |
| Thickness of shell plates | ¾" |
| Thickness of tube plates | ¾" |
| Thickness of furnace tops and fire-boxes | 7/16" |
| Thickness of furnace bottoms | ½" |

Thickness of tubes.... .................No. 9, B. W. G
Thickness of stay tubes........................ ¼"
Stays distance center to center in steam space.................14" × 11" to 12"
Stays distance center to center, fire-box sides................... 6" × 7"
Diameter of 18 long stays in steam space, in 3 rows, 4, 6, and 8....... 1⅞"
Diameter of other stays................. ................. 1¼"

### SUPERHEATER.

Mean diameter shell................ 8' 4⅞"
Mean diameter flue................ 5' 6½"
Length................ 8' 0"
Thickness shell plates................ 9/16"
Thickness flue plates................ 9/16"

The boilers originally carried a certificate from the Board of Trade for a pressure of 60 pounds, which after the repairs above mentioned, was renewed, and a year ago in 1875 the pressure was reduced to 55 pounds.

*General Arrangement of Experiment.*—The experiment was conducted in the following manner. One boiler and the super-heater were connected each to the pumps by a pipe 1 inch in diameter, which worked the hydraulic cranes, whereby a good supply of water was at command up to 800 pounds per square inch; on each a pressure gauge which had been carefully tested previously was fixed, and the experiment was conducted by raising the pressure in both boiler and super-heater successively to 120 pounds, 180 pounds, and 210 pounds, at each of which pressures the boiler and superheater were carefully gauged, the boiler bursting at 230 pounds and the superheater at 245 pounds per square inch. At 120 pounds there was no leakage; at 180 pounds, a little water trickled at some of the longitudinal joints of the shell and at some three-ply rivets at the junction of the tube plates and furnace crown, which had been rather severely acted on by the fire, and had evidently been repaired at some former time. At 210 pounds the leakage in the shell increased, while that in the furnace crowns seemed nearly constant. Up to the time of bursting, the leakage in the shell was not serious, and the only leakage in the furnaces was at the rivets above described. All the rest of the boiler, the furnaces, tubes, fire-boxes, tube-plates, and ends, were perfectly tight and no signs either before or after the boiler burst could be detected in the appearance of the caulking or otherwise that these parts had been subjected to pressure.

*Superheater.*—At 120 pounds there was no leakage, at 180 pounds a little water trickled at the longitudinal joints in the shell, and a few drops leaked slowly in the flue. At 210 pounds the longitudinal joints of the shell leaked very considerably, while the leakage in the flue increased but slightly. At 230 pounds the leakage at these joints in the shell became so great that the pumps could not raise the pressure higher, and one of these joints had to be severely caulked three times before the pressure could be raised to 245 pounds, at which pressure one of the joints opposite to the one which had been caulked gave way.

The fracture of the boiler commenced at 230 pounds, without warning, almost simultaneously in two places: one along a longitudinal seam, the other where the thickness had been much reduced by fitting; it instantly extended through a sheet, and the next seam on one end, and through the front angle iron and tube plate on the other. The impression was that it began on the double rivetted seam. The rivetting was chain, not zigzag. The tensile strength of the plates determined by Mr. Kirkaldy, was from 42,000 to 48,000 pounds. The failure in the superheater was with 37,000 pounds nearly, and in the boiler with 36,500 pounds, tension.

Next in importance to the strength of a boiler is its durability, and it has been said that durability is a function of accessibility; be this as it may, it is certainly extremely important to provide an access to all parts of a boiler both inside and out for the purpose of inspection and cleaning, which sometimes requires the use of a scraper or even a chisel to detach scale. The destruction of a boiler may be caused by corrosion from the outside or from the inside, by overheating caused by scale or by low water, or by the movements and strain caused by changes of temperature.

Internal corrosion is usually caused by chemical action of substances held in solution by the water. The action of oxygen in causing pitting or small spots of rust, is common in boilers which use water from surface condensers, especially when the boiler is laid off frequently and emptied often. Water condensed in heating buildings absorbs much of the air which comes into the pipes whenever the pressure falls below the atmosphere, and such boilers are best kept full of water when not in use.

In boilers supplying steam to engines, there is often trouble from the oil brought over from the cylinder, either from the condenser or from an open heater, and then decomposed by the heat into some one or more of the "fatty acids." The use of only mineral oils in the cylinder will prevent most of this action, while in some cases lubrication of the cylinders has been abandoned altogether, generally for the purpose of preventing foaming in boilers which are crowded to the extent of their capacity.

External corrosion occurs usually at places where leakage of water, or steam running over the plates, is exposed to the action of the fire. It appears that mild English steel is more liable to internal corrosion than wrought-iron. Internal corrosion is often met by hanging in the boilers, under water, pieces of zinc, which, as the weaker metal seem to be attacked in preference to the iron. In cases where the corrosion is from bad water, advice should be sought from a competent chemist. The straining effect due to change of temperature is sometimes enough to produce rupture in a new boiler when this is not properly allowed for in designing. For this reason one head of the Lancashire boiler is attached by an external angle-iron ring, in order that the end may "breath" outward more freely. If we look at the head as strained, a moments consideration will show that it acts as a beam. A given change of length of shell compared with tubes will cause a strain which carries inversely as the square of the distance between the tube and shell, and directly as the thickness of the head; or, if the head be so rigid that the yielding occurs in the shell instead of the head, the result is the same. If the strain exceeds the limit of elasticity and is repeated often enough, the destruction of the sheet by grooving is only a matter of time. The United States law prohibits the placing of tubes in externally fired boilers within 3 inches of the shell, and this is, however, more a precaution against scale filling this up solid, than against expansion. In our opinion the flues of all boilers should not be placed within 4 inches of the shell, when the heads are ½-inch thick, and the distance should be increased with the thickness of the head.

Where there is a tendency to work open the material, or to grooving, as at the head flanges, and where there is no injurious chemical action of the water or deposit, a tough steel flange will resist fracture better than an iron one; but if it once gets broken the steel will go faster than the iron will.

*On Steel.*—Mr. Wm. Boyd concluded from experiments made before the construction of the boiler referred to in a paper read at the Institution of Mechanical Engineers and afterwards published in *Engineering*, pages 310 and 320, for April 19, 1878, for marine use as follows:

1. That steel plates can be had of uniform and reliable material in large quantities.
2. The material is injured nearly one-third, if in punching, the die is $\frac{1}{16}$-inch larger than the punch.
3. It is not hurt by drilling.
4. The quality is restored by annealing.
5. Drilled holes are to be preferred.
6. Especial care is needed in staying flat surfaces. In a paper by Mr. W. Parker, Chief Engineer and Surveyor for Lloyds Register, vide *Engineering*, for June 7, 1878, p. 461, he says of marine steel boilers:

"Now that we have a material that gives us a boiler about 30 per cent. "stronger than an iron boiler of the same scantlings, and as it seems pos- "sible that we may be able in the immediate future to dispense entirely "with longitudinal rivetted seams by having the shells rolled, and as there "has also been a furnace introduced which can work at twice the pressure "of the ordinary plain flue, it does appear to me that we have succeeded "in a great measure in removing the old conditions that have militated "against much higher pressures being obtained, and that we appear to be "now in a position to make a fresh departure in the direction of still "greater pressures. If the improvements which I have indicated, prove, "as I have little doubt they will prove, successful, we shall have gained an "advantage represented in the aggregate by an increase of about 80 or 90 "per cent of the working pressure. In other words we will be able to "work the present form of boiler at 160 pounds or 170 pounds per square "inch, and although the resultant economy will not be so great as that "which attended the increase at one step from 30 pounds to 60 pounds, we "may confidently anticipate that it will be sufficient to give a great im- "petus to steam navigation, advancement in which has lately been so "much retarded by the high consumption of fuel."

In the United States the use of steel is, for stationary boilers, extending rapidly, and there is little else used for the shells and fire-boxes of locomotive boilers, though for stay bolts and rivets, iron is still preferred;

while for boats on the Mississippi River, it is now much used, as allowing higher pressure to be carried.

We give some conclusions drawn for water tube boilers by one of the best authorities, but they embody our views for all boilers.

With regard to water-tube boilers, Mr. Robert Wilson concludes that the points to be attended to are:

1. To keep the joints out of the fire.
2. To protect the furnace tube from cold air when the fire-door is opened.
3. To provide against the delivery of cold feed directly into the furnace tubes.
4. To provide a proper circulation to carry the steam from the surfaces where it is formed.
5. To provide passages of ample size for the upward currents of steam and water, which must be separated from downward currents of water.
6. To provide passages of ample size for the steam and water between the various sections of the boiler in order to equalize the pressure and water level in all.
7. To provide ample surface for the steam to leave the water quietly.
8. To provide a sufficiently large reservoir for the steam to prevent the water being drawn out of its proper place by suddenly opening a steam or safety-valve.
9. To provide against the flame taking a short-cut to the chimney, and impinging against tubes containing steam only.

## CIRCULATION OF WATER.

A very important matter is the movement of the mass of water as a whole in the boiler, caused by the action of the heat and the formation of steam. It has been said by good authority, "that it is no exaggeration to "say that the efficiency and safety of a steam boiler depend as much upon "the efficiency of the water circulation, as they do upon the strength and "disposal of the material of the boiler."

In a plain cylinder boiler we have a furnace hotter at one end than the other, hence the water rises over the fire to a higher level than it has elsewhere, and flows down to the back end along the surface. At the rear end it passes vertically downward and therefore sends, by inertia, the solid matter carried with it to the bottom, or to the flues where a greater incrustation takes place than elsewhere. With the tubular and flue externally fired boilers, the hottest part of the shell, except directly over the fire, is at the highest portion exposed to flame, and as this surface is nearly vertical, a large amount of steam and water rises close to the shell, and if a central space is left between the tubes the bulk of the water will descend in the middle of the boiler This has in one instance of which we have knowledge, been shown by the scour marks of the sand carried in the water.

The importance of a good circulation is evident from a purely theoretic view, for it is well known that the only limit to the amount of heat which can be transferred through an iron plate from one fluid to another is dependent only on the rate at which it can be taken away from and brought to the surfaces; while in the Latta fire engine and the Herreshoff coil boilers, an artificial circulation is maintained by an independent circulating pump. It is of the first importance that in water tubes of a larger diameter than 6 inches, the upward and downward currents of water should be separated, especially when the water is bad; and any heating surface near quiet water is almost sure to form scale. In boilers with water tubes, one end of a tube is usually at a higher level than the other. Small water tubes in large flues, as in the Martin boiler once so largely used in the United States Navy, are now little employed except by Messrs. Shand & Mason, builders of fire engines in London; tubes with one end closed hanging from a sheet above the fire are usually fitted with internal circulating tubes down which the water flows, while the steam formed escapes outside of the circulating tube. The same arrangement is used with the Perkins Surface Condenser to bring up the cold water.

Small cross tubes as in the Martin boiler unless made of brass were found to be subject to rapid external corrosion, when the boilers were frequently laid off, and when made of brass the expense and difficulty of maintaining the joints, and the impossibility of cleaning soot, have prevented their adoption.

## ON HEATING SURFACE.

Rankine gives the following formula and table based upon theoretic grounds as expressing the probable efficiency of a furnace where the ratio of evaporation obtained, $E'$, to the theoretic evaporation of the fuel, $E$, is called the efficiency,

$$\frac{E'}{E} = \frac{BS}{S + AF}$$

where $S$ = square feet of heating surface, and $F$ = pounds of fuel burned per hour. $A = 0.5$ for chimney draft, and $0.3$ for forced draft. $B = 1$ for best connection, $0.917$ for ordinary boilers and chimneys and $0.95$ for ordinary boilers with forced draft. He classes:

I.   For boilers, with best    connection, chimney draft.
II.   "      "     "  ordinary    "         "       "
III.  "      "     "  best        "      forced
IV.   "      "     "  ordinary    "         "

$\frac{E'}{E}$ FOR CLASS.

| $\frac{S}{F}$ | I. | II. | III. | IV. |
|---|---|---|---|---|
| 0.1 | 0.16 | 0.15 | 0.25 | 0.22 |
| 0.25 | 0.33 | 0.31 | 0.45 | 0.43 |
| 0.5 | 0.50 | 0.46 | 0.62 | 0.59 |
| 0.75 | 0.60 | 0.55 | 0.71 | 0.68 |
| 1.0 | 0.66 | 0.61 | 0.77 | 0.73 |
| 0.25 | 0.71 | 0.65 | 0.81 | 0.77 |
| 1.5 | 0.75 | 0.96 | 0.83 | 0.79 |
| 2.0 | 0.80 | 0.73 | 0.87 | 0.83 |
| 2.5 | 0.83 | 0.76 | 0.89 | 0.85 |
| 3.0 | 0.86 | 0.79 | 0.91 | 0.86 |
| 6.0 | 0.92 | 0.84 | 0.95 | 0.90 |
| 9.0 | 0.95 | 0.87 | 0.97 | 0.92 |

D. K. Clark from examination of boiler trials concludes as follows: The water evaporated per square foot of grate per hour, $w$, when the fuel per square foot of grate, $f$, is given is:

$w = a\,r^2 + b\,f$ where $r =$ number, square feet of heating surface per square foot of grate, and $a$ and $b$ are constants with different value for different classes of boilers and fuels, as follows:

| FOR ENGLISH SOFT COAL. | $a =$ | $b =$ |
|---|---|---|
| For stationery boilers............................................. | 0.0222 | 9.56 |
| For marine boilers................................................. | 0.016 | 10.25 |
| For portable boilers............................................... | 0.008 | 8.6 |
| Locomotive, Coal.................................................. | 0.009 | 9.7 |
| Locomotive, Coke................................................. | 0.0178 | 7.94 |

He says the rate of fuel per square foot of grate should not be less than given below, in order to apply the above:

| | $r =$ | | | | | | | | | | | |
|---|---|---|---|---|---|---|---|---|---|---|---|---|
| | 5 | 10 | 15 | 20 | 30 | 40 | 50 | 60 | 70 | 75 | 80 | 90 | 100 |
| Stationary............ | .2 | .7 | 1.7 | 3. | 6.8 | 12.1 | 18.9 | | | | | | |
| Marine................. | .17 | .7 | 1.6 | 2.8 | 6.3 | 11.2 | 17.5 | | | | | | |
| Portable............... | .05 | .2 | .4 | .8 | 1.8 | 3.2 | 5.0 | | | | | | |
| Locomotive, Coal... | .1 | .5 | .7 | 1.3 | 2.9 | 5.2 | 8.1 | 11.7 | 15.9 | 18.3 | 20.8 | 26.3 | 32.5 |
| Locomotive, Coke.. | .1 | .4 | 1.0 | 1.8 | 4.0 | 7.0 | 11.0 | 16 | 21 | 25 | 28 | 36 | 44 |

Reducing to the same basis

$$E' = \frac{E\ B\ S}{S + A\ F}$$

$$E' = b + \frac{a\ r^2}{f} = \frac{f\,b + a\,r^2}{f}$$

$$\frac{E'}{E} = \frac{F\,b + a\ S^2}{F\,E}, \text{ expressions}$$

which are radically unlike and cannot well be harmonized except that in either case it may be seen that the more heating surface per pound of fuel burned the better economic performance, while the latter shows that a reduction of grate area will improve the evaporation, as $r$ and $f$ are increased together.

This is true only within certain limits, as it is difficult to burn more than a certain quantity of coal per square foot of grate per hour without increase in the draft.

Of these two formulæ, that of Clark seems more nearly to apply to the experimental data, although his constants meet with great change. Thus from the experiments upon the Wabash Railway with coal from Central Illinois, Clark's constants become for locomotives

$$b = 4.5 \qquad a = 0.05$$

and $E' = 4.5 + \dfrac{r^2}{20\,f}$

The proportions of heating surface per square foot of grate is usually taken as the governing element, and to Chief Engineer Isherwood is due the first demonstration, that boilers may be made too large for the purpose of economic evaporation. Lately Mr. Alfred Blechynden has made an extended mathematical investigation of the subject, and he concludes that for land boilers an efficiency of 0.75 is desirable, and for yachts and boats intended for speed without counting freight, that with fuel to be had at intervals the desirable efficiencies are as given below in the tables:

DESIRABLE EFFICIENCIES FOR VARIOUS PRESSURES, AND TIMES BETWEEN COALING, FOR ORDINARY MARINE BOILERS—ORDINARY DRAFT.

| PRESSURE. | NUMBER OF DAYS FOR WHICH COAL IS CARRIED. | | | | | | | | | |
|---|---|---|---|---|---|---|---|---|---|---|
|  | ¼ | ½ | ¾ | 1 | 2 | 3 | 4 | 5 | 10 | 20 |
| 60 | .247 | .314 | .358 | .390 | .408 | .514 | .547 | .571 | .642 | .704 |
| 75 | .237 | .303 | .346 | .377 | .456 | .502 | .535 | .559 | .631 | .695 |
| 90 | .229 | .293 | .335 | .366 | .444 | .491 | .524 | .548 | .625 | .686 |
| 105 | .221 | .284 | .325 | .356 | .434 | .481 | .513 | .538 | .612 | .678 |
| 120 | .214 | .276 | .317 | .348 | .425 | .472 | .503 | .529 | .604 | .671 |
| 135 | .208 | .269 | .309 | .339 | .416 | .463 | .495 | .521 | .596 | .668 |
| 150 | .203 | .262 | .302 | .332 | .408 | .455 | .487 | .513 | .589 | .658 |

## STEAM JET DRAFT.

| PRESSURE. | NUMBER OF DAYS FOR WHICH COAL IS CARRIED. | | | | | | | | | |
|---|---|---|---|---|---|---|---|---|---|---|
| | ¼ | ½ | ¾ | 1 | 2 | 3 | 4 | 5 | 10 | 20 |
| 60 | .307 | .382 | .430 | .464 | .545 | .592 | .623 | .646 | .712 | .770 |
| 75 | .295 | .370 | .417 | .451 | .532 | .579 | .612 | .635 | .703 | .762 |
| 90 | .285 | .359 | .405 | .439 | .521 | .568 | .601 | .625 | .694 | .754 |
| 105 | .276 | .349 | .395 | .429 | .510 | .558 | .590 | .615 | .685 | .747 |
| 120 | .268 | .340 | .385 | .419 | .500 | .548 | .581 | .606 | .678 | .740 |
| 135 | .261 | .332 | .376 | .410 | .492 | .539 | .573 | .598 | .670 | .734 |
| 150 | .255 | .324 | .369 | .402 | .484 | .531 | .565 | .590 | .664 | .728 |

Mr. Blechynden's expressions become too complex when the value of freight is taken into account, and we give a simpler if less exact investigation as an example of the method to be pursued.

Ordinary boilers weigh on board say 40 pounds per square foot of heating surface, costing not far from $4. Allowing for interest on first cost and repairs, together, 80 cents per annum, or 20 per cent. Freight per ton per working year may be assumed as $50, or $1 per annum per square foot of heating surface, making an annual cost per square foot of heating surface of $1.80.

Suppose fuel costs $4 per ton and the boat is under weigh 50 days per year, then for each pound of fuel per hour burned there is expended $2 per year and the freight values are, if fuel is taken every day, $1.20. If freight can be taken on board every hour the average values may be taken as half that given or 60 cents in place of $1.20.

Suppose that from the engine power desired we know that we have to evaporate 20,000 pounds of water per hour we make out the following table by the aid of the Rankine formula reversed:

$$S = \frac{E' A F}{E B - E'} \text{ taking } A = \tfrac{1}{2},\ B = 0.9$$

we have then $S = \dfrac{10000}{13.0 - E'}$

| Evaporation in pounds of water per pound of coal. | No. square feet of heating surface required. | Cost in dollars per year of interest and freight on boiler. | Cost of coal for fifty days in steam. | No. of pounds of coal burned per hour. | Sum of columns three and five. | Cost for freight for one day's coal. | Sum of columns six and seven. |
|---|---|---|---|---|---|---|---|
| 3 | 1000 | 1800 | 6667 | 13333 | 15133 | 4000 | 19133 |
| 4 | 1111 | 2000 | 5000 | 10000 | 12000 | 3000 | 15000 |
| 5 | 1250 | 2250 | 4000 | 8000 | 10250 | 2400 | 12650 |
| 6 | 1428 | 2580 | 3333 | 6667 | 9247 | 2000 | 11247 |
| 7 | 1667 | 3000 | 2860 | 5718 | 8718 | 1715 | 10433 |
| 8 | 2000 | 3600 | 2500 | 5000 | 8600 | 1500 | 10100 |
| 9 | 2500 | 4500 | 2222 | 4444 | 8944 | 1333 | 10277 |
| 10 | 3333 | 6000 | 2000 | 4000 | 10000 | 1200 | 11200 |
| 11 | 5000 | 9000 | 1818 | 3636 | 12636 | 1111 | 13747 |
| 12 | 10000 | 18000 | 1667 | 3333 | 21333 | 1000 | 22333 |

If the evaporative efficiency of the fuel is not so high $E$ is lessened and we have different values. If $E = 12$ we have $S\dfrac{10000}{11-E}$ and our table with columns:

| Evaporation in pounds of water per pound of coal. | No. square feet of heating surface required. | Cost in dollars per year of interest and freight on boiler. | Cost of coal for fifty days in steam. | No. of pounds of coal burned per hour. | Sum of columns three and five. | Cost for freight for one day's coal. | Sum of columns six and seven. |
|---|---|---|---|---|---|---|---|
| 5  | 1667  | 3000  | 4000 | 8000 | 11000 | 2400 | 13400 |
| 6  | 2000  | 3600  | 3333 | 6667 | 10267 | 2000 | 12267 |
| 7  | 2500  | 4500  | 2860 | 5720 | 10220 | 1715 | 11935 |
| 8  | 3333  | 6000  | 2500 | 5000 | 11000 | 1500 | 12500 |
| 9  | 5000  | 9000  | 2222 | 4444 | 13444 | 1333 | 14777 |
| 10 | 10000 | 18000 | 2000 | 4000 | 22000 | 1200 | 23200 |

If the type of boiler used weighs more or less than 40 pounds per square foot of heating surface, column three may be varied accordingly. If the boiler is to be in full steam for more or less than 50 days in the year or fuel costs more or less than $4 per ton, column five is to be varied accordingly. Column six will do for land boilers. If the coal is be carried for more or less than one day, column seven is to be varied and from the sum we easily find the proper heating surface. The real quantity of water to be evaporated is not essential as all the quantities in the table will vary therewith also.

If instead of freight charges we take the value of land occupied, and if fuel has to be stored, the room it occupies, we shall see that there is not perhaps, so great a difference between steamboat and stationary boilers as was supposed by Mr. Blechynden. And from the table we can see that the real element is the time per year the boilers are to be at work. In a flour mill running 144 hours per week for instance, a high economic performance is very desirable, but for a water works with storage reservoir where one day's pumping in a week will keep up the supply and fuel is obtained as wanted, an evaporation over seven or eight is not desirable until the work is increased. The Western River Steamboat Boiler is thus seen to be admirably adapted to the conditions in which it works, while for an ocean line steamer it would be out of place.

The kind of boiler to be selected for any particular work or locality depends upon many things, and it is seldom advisable to make very wide departures from the common practice around the given locality. Thus with well water, locomotives in Central Ohio sometimes have to have their tubes renewed every six months on account of scale and consequent burning of ends. It would not be advisable to put in a locomotive type boiler for a flour mill in this region and you will find few tubular boilers there. The flue return, externally fired type, or the French type is suitable, although the latter have been little used in this country. The externally fired return tube type is common with good water, while the Cornish and Lancashire are almost unknown and from their cost and the higher pressures used are not likely to become favorites with the United States.

For hard steady work where land is worth little the single cylinder type has many good points. For high pressures the water tube type is in many places a favorite. The internally fired boilers of the locomotive type usually give high evaporative results but are hard to keep clean.

The externally fired return tubular with central gangway seems to combine more good features for less money than many other types and without the gangway is probably more used in the United States than any other. With two to five flues is the standard river practice for the Mississippi and tributaries.

The choice of marine types is governed by the room at disposal as to whether a single or double fire room can be had.

A properly constructed boiler requires a very great internal pressure to burst it. A 36-inch shell ¼ inch thick and 10 feet long, with ordinary rivetting, has 18 square inches of metal, which requires at least 50,000 pounds per square inch, tensile strength, in all 450 tons, or 416 pounds per square inch to burst it; and by the United States Law is allowed 125 pounds per square inch pressure; or on a tow boat on the Mississippi River a pressure of 175 pounds to the square inch.

With this margin of strength the occurrence of violent explosions has been taken to justify the hypothesis of some violent internal action, and as there are probably more cases of quiet failure than of violent explosion, there has seemed a certain probability in this view. A simple rupture attended by the loss of steam and water under ordinary working can occur only from a purely local failure of a seam, or rivet, either from original defect in material or manufacture, or from subsequent injury, and such cases seldom attract public attention.

The failure of boilers, whether violent or quiet, has been attributed to steadily accumulated pressure; steam formed from sudden contact of water with red hot metal; electrical action; the decomposition of water or steam into hydrogen and oxygen, etc. Let us first consider overheating. Although it is possible for boilers to be exploded in consequence of the formation of steam by the contact of water with very hot plates, yet overheating cannot be taken as the only cause, or even the general one, of explosions, for there is too much evidence that boilers do explode with plenty of water in them. Burnt iron is easily recognized and its absence is good evidence of a sufficient quantity of water.

Admitting the case of overheating, it is doubtful whether the formation of steam would cause an explosion, for the actual quantity of heat which the metal can hold is not capable of very much work in the way of generating steam, as may be best seen by an example computed by the aid of our table of the properties of steam. Suppose a boiler with a water space of 100 cubic feet and a steam space of 40 cubic feet becomes short of water, and by reducing the water to 90 cubic feet, increasing of course the steam space to 50 cubic feet, a surface of 100 square feet of ¼-inch plate weighing

## DESIGN AND CONSTRUCTION CONTINUED, ETC.   163

1,000 pounds is uncovered; that the steam pressure is 100 pounds per square inch, and that the iron plates uncovered rise to 1,000° F. hotter than the steam and water, or to 1,338° F., what will be the consequence of pumping in 10 cubic feet of water at 100° F. in five minutes?

The condition of things at first is: 90 cubic feet of water at say 60 pounds, is 5,400 pounds of water × 308 units = 1663200 units; 50 cubic feet steam × density 0.263 = 13.15 pounds; 13.15 pounds × 1184 units = 15570; total 1678770. The heat stored in the hot iron is at best 1000 pounds × 1000° × 0.111 the specific heat, or say 111000 units. If this were all to be put into the steam it would superheat it to a very great degree and undoubtedly raise the pressure beyond the bursting pressure. Let us examine this by steps, first supposing no fresh water to have been added, and that the pressure has raised to 110 pounds. At that time there will be present as steam

|   | Units. |
|---|---|
| 50 × 0.284 = 14.2 pounds, and its heat at 1186 units is.................................. | 16841 |
| The 5,400 pounds of water is now at 344° F. and the heat is 315 units.............. | 1701000 |
|   | 1717841 |
| Formerly........................................................................................ | 1678770 |
| So there has been added from the iron.................................................... | 39071 |

If, however, the 10 cubic feet of water have all been pumped in, the account will be very different. At 110 pounds we have 40 cubic feet of steam,

|   | Units. |
|---|---|
| 40 × 0.284 = 11.36 pounds at 1186 units will give....................................... | 13473 |
| 6000 pounds of water at 315........................................................... | 1890000 |
|   | 1903473 |
| Deduct 600 pounds of water feed from 32° to 100°, not furnished 600 × 68..... | 40800 |
|   | 1862673 |
| Originally....................................................................................... | 1678770 |
|   | 183903 |
|   | 111000 |
|   | 72900 |

being in excess of that stored in the iron, say 70 per cent, showing that a rise of pressure in five minutes of over six pounds to seven pounds is not at all likely. It has, however, been taken that there was no heat from the outside coming in, and this is not likely to happen; still this also fairly represents the case when an engine is at work and the fire is in such condition that the regular supply of heat is just enough to furnish steam for the engine. Suppose the engine is using 1,000 horse-power with 3,600 pounds of water an hour, in five minutes it will use 300 pounds, and the regular heat from the fire is to be estimated 1184 — 68 units = 1116 units × 300 pounds = 334800, which if added to our 111000 stored makes up 449000 units or enough to raise the pressure to say 140 pounds if the boiler has no outlet. Let us examine it. At 140 pounds we have

```
                                                                    Units.
40 × 0.348 = 13.92 pounds of steam at 1191 units................... 16589
6000 pounds of water at 331 units..................................1986000
                                                                  ────────
                                                                  2002589
Deduct as before not furnished.....................................  40800
                                                                  ────────
                                                                  1961789
                                                                  1678770
                                                                  ────────
                                                                   283019
```

which is less than the supply 449000 and the pressure will rise higher. Let us try it at 180 pounds:

```
                                                                    Units.
Steam 40 × 0.433 = 17.32 pounds of steam at 1197................... 20733
6000 pounds of water at 351........................................2106000
                                                                  ────────
                                                                  2126733
      Less as before...............................................  40800
                                                                  ────────
                                                                  2085923
      Originally...................................................1678770
                                                                  ────────
                                                                   407163
```

Being nearly the amount at hand, the pressure will go above this point a few pounds, and is undoubtedly a dangerous pressure on an old or weakened boiler, but could not be reached with a proper and effective safety-valve in operation.

We will next consider the question of electrical action which is often brought forward, supported by the Armstrong generator. Faraday found in his examination of Armstrong's apparatus that the boiler had to be insulated, the steam wet and the nozzles of boxwood. He concluded that the production of electricity was not due to any change of state of the liquid in the boiler, and that the same results could be obtained by moist compressed air. Without going further we can say no one has shown how a boiler can be exploded even by any quantity of electricity even if it were there, as with the excellent conduction of the quantities of iron around it anything like a sudden discharge would be impossible. We will leave this theory to those who prefer mystery, and who are ready to see in electric action a cause rather than an effect of things not otherwise explained.

The decomposition of steam into hydrogen by the absorption of oxygen by red hot iron is an experiment which requires very different conditions from the ordinary working of a boiler, and even granted that we had a boiler full of water and hydrogen gas under pressure we do not see how any amount of air containing the fresh oxygen necessary for any explosion could be introduced. We might as well expect an ordinary gas holder to explode from the same cause. We conclude then that decomposition can not occur in the ordinary working of a boiler, and if it did no explosion would follow unless mixed with oxygen, and if so mixed in the presence of steam no combustion could take place, and if no steam were present we might expect only a quiet combustion as the air was forced in, the ignition coming from some red hot plate. A piece of burnt iron is evidence of shortness of water and of a weakening action upon the metal without any gas explosion being required. We may now pass to what is called over-pressure.

Any pressure greater than the safe working pressure upon a boiler is an over-pressure, and the result may be rupture with or without an explosion. The action of corrosion in reducing the thickness and of change of form by change of temperature in producing strain upon the metal, tend to reduce the bursting pressure and in many explosions the plates are found not more than $\frac{1}{16}$ inch in thickness. Mr. Robert Wilson states that it is possible to make a boiler which would be tight with 100 pounds hydrostatic pressure but that would explode with 30 pounds steam pressure, or which in other words would almost tear itself to pieces when a portion was heated to 300° F., or to a greater temperature than some other portion.

The operations which occur very rapidly after each other during an explosion are probably as follows:

1. The rupture under a pressure not much greater than the working pressure of some defective portion of the boiler.
2. The extension of this rent through some adjacent portion of the boiler, owing to a transference of strain, or by otherwise yielding to a shock.
3. The lowering of pressure caused by the escape of fluid from the boiler and the sudden generation of steam at the lower pressure, causing a following up of the pressure upon the moving pieces thereby causing the *violent* characteristics which may take place.
4. The ejection in whole or part of the water contained in the boiler mixed with steam formed from the hot water, of which, of course, only a small portion can be vaporized by the heat contained in the water, as 965 units are required to boil water from and at 112° while at 140 pounds the temperature is 361° or 149° units higher than 212°—965 divided by 149 is $6\frac{1}{2}$ nearly. That is to say, of every $6\frac{1}{2}$ pounds of water in the boiler one pound will be thrown into steam at atmospheric pressure, while $5\frac{1}{2}$ pounds will remain as water either in the boiler or more likely scattered in the air mixed with the steam resulting. In general all our knowledge of boiler explosions goes to show that in most cases the explosion results from some defect either original or produced, either visible or concealed in the materials, workmanship, or design and construction of the boiler. Probably less than one per cent. of the boilers made explode, but many more are ready to fail either quietly or violently from causes which may be easily discovered by competent inspection.

Some valuable experiments were made in 1870 at Sandy Hook, by Mr. Francis B. Stevens, of Hoboken, at the expense of the united Railroads of New Jersey. Several old boilers had been taken out of steamers belonging to the united companies and were burst by hydrostatic pressure and repaired and again burst several times, finally leaving them much stronger than when taken from the boats. Professor R. H. Thurston, of the Stevens' Institute of Technology, gives an account of the experiments, which we here sum up. The boilers after this process of finding out the weak places were set up at Sandy Hook at the entrance of New York Harbor, and three of the boilers were burst as follows:

The first was a fire box boiler 28 feet long with shell 6 feet 6 inches diameter and barrel 20 feet 4 inches long. The two furnaces were 7 feet long with flat arches and with 4 inch water space all around and water leg between the furnaces. The products of combustion passed through a throat to a combustion chamber 19 inches long, and then through 10 flues of which two were 16 inches and the other eight 9 inches in diameter and 15 feet 9 inches long. The back connection was 32 inches in length with a 4 inch water space behind it. The return tubes above were 12 in number and $8\frac{1}{2}$ inches in diameter, and 22 feet long to the smoke connection, which formed the base of the stack 2 feet 8 inches in diameter surrounded by a steam chimney 4 feet in diameter and 10 feet 10 inches above the shell of the boiler. The grate area was $38\frac{1}{2}$ square feet and the total heating surface was 1,350 square feet; the flat surfaces were stayed with screw bolts at 7 inch centers. This boiler was one of a pair built in 1856 and had been 13 years in service, the last inspection certificate allowed 40 pounds pressure. In September of 1871 this boiler had been tested by hydrostatic pressure to 66 pounds pressure, when one of the stay bolts pulled through; after repairs it was tested to 82 pounds, and afterwards steam of 60 pounds had been made in it. On November 22, 1871, the water standing 12 inches over the flues a heavy wood fire was made in the furnaces and steam raised to 50 pounds when the party retired to the gauges, a distance of 250 feet behind screens; at 90 pounds leakage occurred, and at 93 pounds the connection between the shell and dome or steam chimney failed on the top of the shell and the steam passing off the pressure failed, and no explosion took place.

The next experiment was made with a small flat stayed box 6 feet long, 4 feet high and 4 inches thick made of two sheets of $\frac{7}{16}$ inch "best flange fire box iron" from the Abbot Iron Company. The edges were rivetted through a bar with $\frac{3}{4}$ inch rivets 2 inch centers, the stay bolts were $8\frac{3}{4}$ inches long, $9\frac{3}{16}$ inch centers with the ends slightly rivetted over. It had carried 138 pounds hydrostatic pressure without difficulty. It was set on edge in brickwork and was about $\frac{2}{3}$ full of water. The pressure rose in 33 minutes from 0 to 167 pounds when a violent explosion took place, the bolts pulling through the plates without injury to the threads.

The next day a rectangular boiler was tested, 15 feet 5 inches long, 12 feet 2 inches wide and 8 feet 6 inches high, for half the length from the front, and a foot less for the remainder. The furnace and combustion chamber extended 14 feet 8 inches to the rear of the back connection and was 11 feet 5 inches wide. The back connection was 18 inches in length and the tubes were 12 feet in length to the smoke connection, 2 inches in diameter and 384 in number. The water legs were stayed by 1 inch screw bolts 12 inches by 8-inch centers. The sides and ends by ties, $1\frac{3}{8}$ rods 28 inches, by 12 inch centers. The furnace crown and top by "crow foot bars" $\frac{1}{2}$ inch by 2 inches, 12 inch by 17 inch centers. The shell was of No. 3 iron single rivetted and there was a dome in the middle of the top 6 feet in diameter and 8 feet 8 inches in height. This boiler was built in 1845 and after a service of 25 years was taken out. The last inspection allowed a pressure of 30 pounds. Forty-two pounds of water pressure broke a crown brace and at 60 pounds 12 of these braces failed; after repairs it carried 59 pounds of water and 45 pounds of steam safely. The furnace was filled with as much wood as would burn freely and the pressure rose in 13 minutes from $29\frac{1}{2}$ pounds to $53\frac{1}{2}$ pounds, when it burst with a violent explosion, the boiler being torn in many places, the dome rising to above 200 feet in the air and to a distance of 450 feet.

The conclusions reached by Professor Thurston are for stay bolts the greatest distance from centre to centre in inches should be 365 times the thickness of plates divided by the square root of the pressure. The latter should be multiplied by the factor of safety desired, which in this case should not be less than 6. (The United States Navy use 8.) The diameter of screw bolts should be twice the thickness of the sheet with $\frac{1}{4}$ inch added. The conclusions drawn are as follows:

1. A violent explosion may take place in a boiler when there is plenty of water in it.

2. A moderate pressure of steam may produce a terrific explosion when there is plenty of water.

3. That a boiler may explode under steam at a less pressure than it has stood from water pressure without apparent injury.

The above conclusions are not new but this was the first experimental demonstration of them, though the first and third had been proved by facts.

We add to these a fourth, which is: A rupture will be followed by relief of pressure with or without explosion as the fracture is extended or restricted.

Implicit confidence cannot be placed in the hydrostatic pressure test, but it should be followed with careful inspection assisted by the sound of a blow struck with a light hammer. Many boilers have exploded when corrosion has reduced the metal to such a thickness that a smart blow from a round-ended hammer would have gone through the sheet, while the boiler has shown tight when under hydrostatic pressure.

## CHAPTER VIII.

### MISCELLANEOUS BOILERS.—CHOICE OF BOILER FITTINGS AND APPURTENANCES.

As we have seen, each type of boiler has its distinctive features. Those with large grates are suited for a class of fuel which requires air in close contact, and room for evaporation of the water which is contained therein—the grate bars are often replaced by a plate with or without perforations. Sawdust and fine chips or shavings are burned on a close plate. Wet fuel requires room and is usually some kind of refuse which is to be got for low cost. Crushed sugar cane refuse, or "bagasse," was probably the first example of this kind, for the saw-mill furnaces were more generally fired with edgings and slabs than with sawdust. Spent tanbark has been used in the tanneries, and much time and money has been spent in improvement.

The most common refuse fuel is that from the thrashing machine—being the chaff after removal of the grain. These machines are usually driven by a portable engine, and the fuel is often mixed.

Refuse coal is also a fuel which engages a great deal of attention, and thus very many attempts have been made to devise a furnace for universal use. With soft coal little trouble is met with except from the earthy matter and clinkers, the latter are a nuisance and require some trouble in management. Coke dust, or "breeze," burns well with a forced draft, and refuse anthracite, if burned with a moderate rate of combustion if not disturbed, burns well for a certain time. The removal of ash and earthy matter makes it difficult to keep up the fire.

Large flues are required for fuel which has an excess of hydro-carbon in order to keep the gas from getting chilled, and to give time for combustion. Small flues require constant care to keep them from silting or sooting up. With a strong draft less trouble is experienced, but with good proportions, and an anthracite fire, flues need not be swept oftener than once a week, while with the poorer kinds of western coal, and the same boiler, once in twenty-four hours would not be excessive care.

The water circulation, steam and water room have all their influence in affecting the durability of the boiler, and many forms of boilers which have been successful in a few isolated cases, or even in many cases, confined to one locality would, if erected in different circumstances, prove an entire failure.

The quality of the water,—the character of material held in suspension and solution, is one of the most important factors in making choice of a boiler, as also the usage, whether constant or intermittent, and whether uniform or varied, while it is at work. The opportunities afforded for examination and cleaning are also to be considered.

It is thus evident that we shall meet many cases where a particular type of boiler has been successful in isolated examples, and even where many boilers of a given type are working to satisfaction, within a definite boundary, which would not bear transportation beyond that boundary. Heretofore we have confined ourselves to the general types of boilers and with enough of comment to show the special fitness for certain uses.

We have seen that with regard to evaporation that both in economy and capacity the results were governed entirely by the magnitude of the heating surface, the amount and intensity of the fuel burned and the quality and kind of fuel, and not to any extent by the disposition of the material of the boiler and furnace further than it affects the more or less complete process of combustion. And that so long as ordinary care is taken to maintain the combustion arrangements in a state of efficiency one boiler is as good as another and a choice must be made on the other considerations.

1.—*The Fuel.*—If anthracite is used, a grate area for burning from 6 to 20 pounds of coal per square foot per hour must be provided for a stationary boiler, and from 12 to 30 pounds for a marine boiler. For a locomotive from 30 to 100 pounds—the former being a rate found with slack and waste on the Philadelphia & Reading Road, while the latter is reached by passenger engines. About 300 pounds per square foot of grate seems to be a limit to the duration of the fire, at the expiration of which it has to be renewed; but with anthracite renewal means rebuilding after that time, and if fuel is added during the time in question its effect is commonly to put out what has been burning without getting ignited itself. Any attempt to clean the fire with a rake or slice bar usually puts it out. The more there is than 20 to 25 pounds per square foot of grate per hour burning the more certain are these results.

So much for the grate required for anthracite. The size of flues, tubes, and arrangement of furnace is a matter of indifference, save that the area through the flues is usually taken as $\frac{1}{7}$ to $\frac{1}{8}$ of the grate; with a forced draft this may be diminished without interfering with the capacity while the economy may be increased, or if it be enlarged, the economy remaining the same, the capacity may be increased.

With good bituminous coal the flues are often slightly increased, and more room is required in the furnace,—the rates of combustion may be considered as 50 per cent. in excess of those for anthracite. As the quality of the fuel deteriorates, more room and more grate are required. Mixtures of fuel often produce excellent results—anthracite and bituminous coal, coal and wood, soft and hard wood, etc. In any case care must be taken with the arrangement of the furnace in such a manner that no portion of the metal of the boiler shall become locally much hotter than the balance of the boiler, to avoid the straining and risk of burning the metal produced thereby.

2.—*The Quality of the Water.*—Many substances held in solution or suspension in hot water are deleterious to the metal of the boiler. Some are actually corrosive, some are merely productive of scale which may hide

## OGLE'S BOILER.

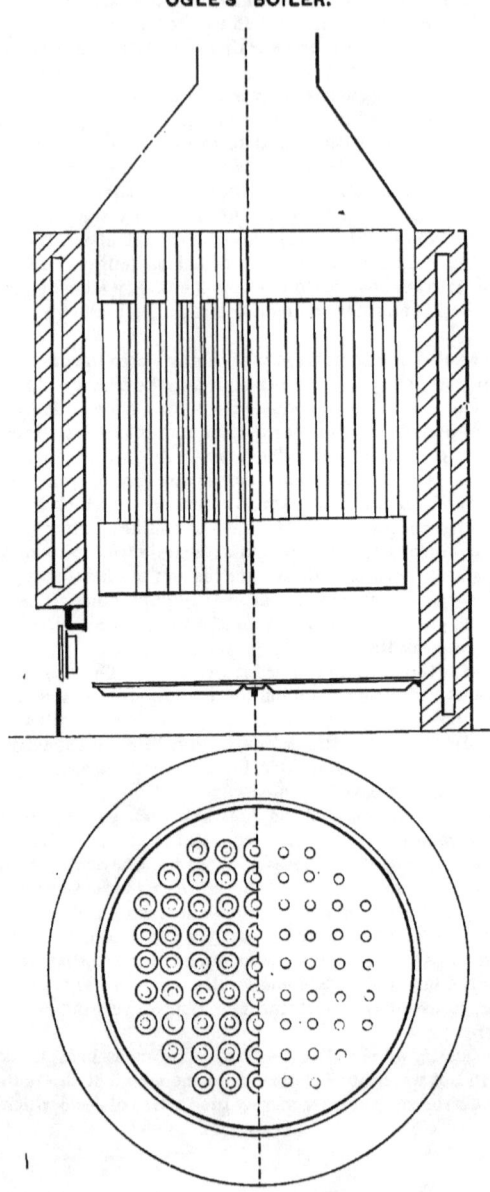

corrosion or may, if of sufficient thickness, cause the metal of the boiler to become overheated. Some substances which are soluble in water at ordinary temperatures and even at 212° F. are insoluble in water at higher temperatures and are, of course, precipitated in fine powder, this powder falls to the bottom in quiet water, but is kept in suspension and rises to the surface in places where it is sufficiently agitated. New combinations are often formed with the mere dirt which has been there all the time in mechanical suspension. Oil and grease returned from the engines by an open-heater or a condenser is a new element, especially at high temperarures. Sulphate of lime is actually less soluble in water of high temperature, and at 50 pounds pressure above the atmosphere is entirely insoluble. Carbonate of lime held in solution in water containing carbonic acid and bi-carbonate of lime deposit carbonate of lime as the temperature of the water rises by driving out the carbonic acid and by separating into carbonate of lime.

These two substances form the basis of most kinds of scale,—a third being simply earthy or sandy matter met with in the water. If the ebullition be strong enough to keep most of this deposit in suspension at or near the top of the water, it may be removed by a surface blower; or if a closed feed heater be used the feed water may be raised by the use of live steam nearly to the temperature of the boiler and the deposit allowed to settle to the bottom or other surface of the heater and removed by blowing or washing out. Care should be taken to wash out with warm water if possible.

It would seem at first sight that distilled water as the return water from a building heated by steam or by a surface condenser would be exactly what was required, but in either case the air which will enter the pipes when the pressure falls below the atmosphere induces pitting, and unless the cylinder oil be mineral, even at ordinary pressures, we find that it decomposes and fatty acids are formed, which act very rapidly.

With the purification of water before usage we have little to do. By the addition of quicklime many evils are removed, but in most cases nothing can be done. We have then after considering the water to determine what may be done. If the water contain only a small portion of impurity, which will form scale, almost any imagined form of boiler may be successfully used.

We have known instances where locomotives fed with surface water in rocky districts in New England have run for twenty years without having a tube removed, while in central Ohio there are regions in which at the end of six months half the tubes would have to be renewed if care were not taken in washing out and cleaning, and it used to be said that the water of Bitter Creek would cause an engine to leak in a fortnight when the Union Pacific R. R. first reached that region.

Among the forms of boilers which can be used with only the best water, is one first invented by Ogle and afterwards modified by Prosser. Two sets of tubes were used—a long and short—and four tube plates, the tube which reached from the first to the fourth tube sheet passed through the short

## THE BELLEVILLE (FRANCE) BOILER.

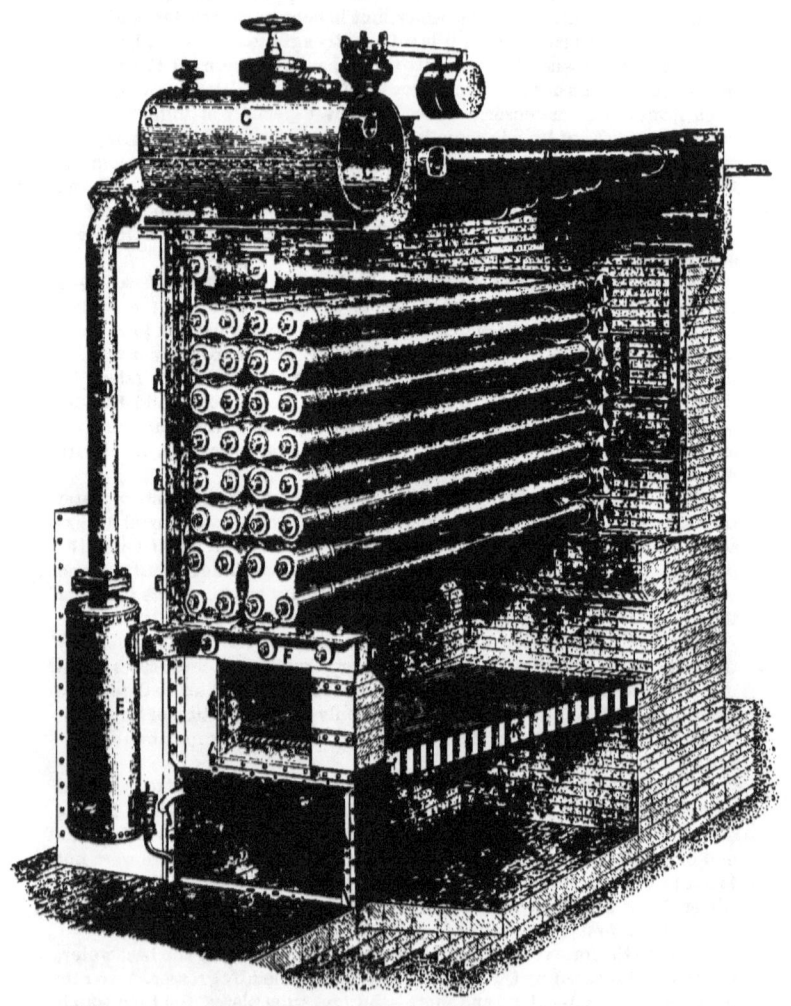

tube which reached from the second to the third tube plate, thus passing tube within tube a chamber was formed by a shell from the first to the second tube sheet, and another from the third to the fourth. The water and steam occupied the space in these two chambers, and the annular spaces between the short and long tubes, provision being made for the descent of water from the upper to the lower chamber by tubes without internal ones. The fire was underneath in a furnace chamber, the gas traversed the inside of the long tubes vertically to a connection chamber, and also passed outside the short tubes and around the two chambers to the smoke connection above. When this arrangement was set in brickwork every bit of surface became heating surface, and by carrying a low water line dry steam was taken from the upper chamber in spite of the rapid circulation which was set up in the annular spaces. With any grease or dirt forming scum, this boiler inevitably foamed, and no way existed for examining or cleaning the inside of the boiler unless made of excessive size; the outside of the short set and inside of the long set of tubes could, however, be considered as fairly accessible. This form reappeared years ago in New York where it was modified by combination with a fire tube and water tube addition and was used for heating a building. Nothing but distilled water was fed into the boiler.

This class has since appeared with further modification. The second and third tube sheets have been united by an iron shell and connection made between the chamber, thus formed, and the steam space in the chamber above it. The products of combustion pass only up through the long tubes, while the middle chamber is supposed, while acting as a drum, to also act as a submerged drying chamber. The complexity, weight for a given heating surface, and difficulty of inspection appear to be all increased by this arrangement.

The various forms of pipe boilers, were first introduced by Jacob Perkins, very many years ago. We may note two types at this day:

The continuous line of pipe in which the water is forced at one end and from which steam and water issue at the other. If the pipe is exposed near the outlet to a great heat an excess of water must be pumped through the pipe to keep it from burning; if on the other hand the water is delivered to the hot end of the pipe no excess of water is required. In the latter case any impurity left by the water after evaporation will be deposited on the inside of the pipe. With excess of feed water it may be carried through the small pipe into a receiver beyond. There may be one line of pipe only, as in Perkins', Elder's and Herreshoff's, or more than one parallel line, as in Benson's, Lattas' and the smaller Belleville forms.

The excess of feed water carried through the pipe into the receiver may be blown off together with the sediment by a bottom or surface blow-off attached to the receiver, or the surplus water may be blown into the feed supply, thereby warming it; in either case being supplied to the boiler through the ordinary feed pump along with the feed water, or it may be taken from the receiver itself and circulated through the pipe returning to the receiver, in which case a separate pump has to be provided.

174    STEAM MAKING; OR, BOILER PRACTICE.

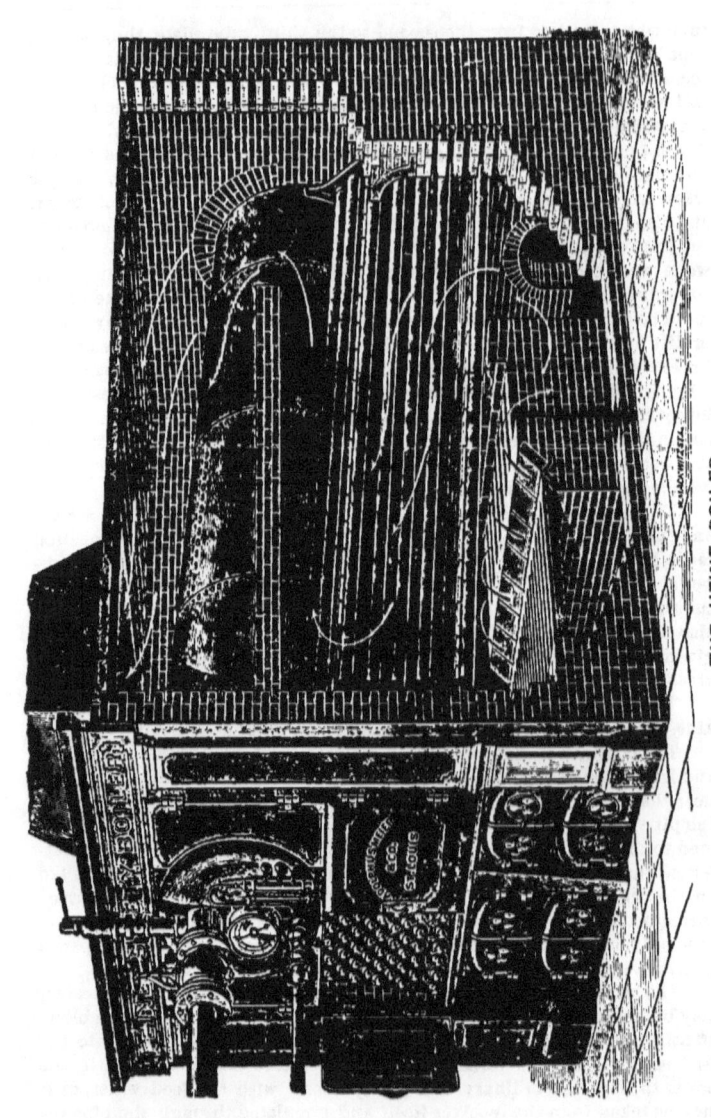

THE HEINE BOILER.

# MISCELLANEOUS BOILERS, ETC.

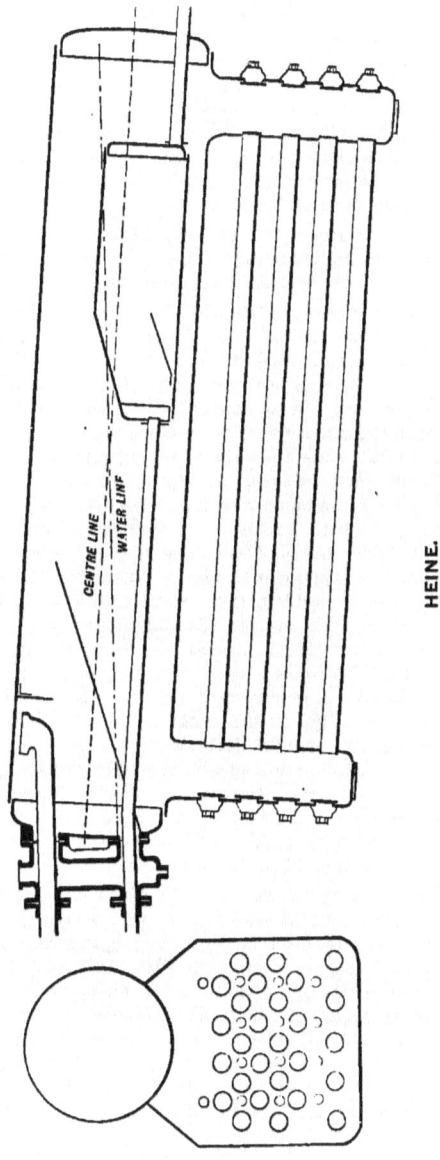

HEINE.

The form of the line of pipe has received many changes. Perkins' has a cylindrical spiral; with Elder a conic spiral; with Herreshoff a combination of cylindrical, conic and flat spirals. With Benson and Belleville a series of nearly horizontal pipes, one over the other, connected at alternate ends. The form of receiver used was a vertical cylinder by Elder, Benson and Herreshoff, while Belleville uses a horizontal cylinder for a steam drum connected with a vertical cylinder used as a mud-drum, from which the blow-offs are led. Latta uses for his receiver a vertical annular space between two shells, the furnace and coils of pipe being inside the inner one.

Benson, Latta, and Herreshoff use an independent circulating pump. Belleville in the small boilers blows off directly, while Elder proposed as a circulator a small screw propeller driven from outside.

With good water the cost of blowing away the surplus feed is small as this water has not been evaporated, but in some cases a surface condenser even without an air pump is used in order to secure a supply of good water, for example at sea with small boats and engines. The cost in fuel of blowing off into this reservoir is, of course, less.. The quantity of excess required varies with the intensity of the combustion, and appears not to exceed one-half more than the feed, the waste of heat is then not likely to exceed 12 per cent. when the surplus is blown away, and 8 per cent. when the feed is returned to the balance raising it to a high temperature. With a circulating pump very little heat is lost. With water containing only ordinary impurities these classes of boilers may be used advantageously, the excess being thrown away; but with sandy and muddy water it is always found that the water brought in the feed pipe to a boiler begins to deposit solid matter outside the boiler, and to fill up the pipe until the opening is just enough to supply what is needed. As feed pipes are usually many times the size required for this purpose (a 2-inch pipe often closing up to less than $\frac{1}{4}$ of an inch) it is readily seen that if the feed be introduced, as it should be theoretically, at the coolest part of the pipe, this may easily happen to the coil pipe itself, and then that some accidentally large portion of mud accumulating during a short cessation of work, would either close, or so nearly close the already contracted passage, and the portion of pipe exposed to the greatest heat of the furnace would burn before the obstacle would wash through. A boiler containing a large body of water is not so soon subjected to the risk of burning.

A second class of pipe boilers is one in which the circulation is provided for only by gravity. As examples of this we have the Babcock and Wilcox (already illustrated); the larger Belleville, and the Heine, Root and Firmenich, which we give. The latter has for its prototype a combination of curved pipes, each a semi-circle, uniting two straight pipes of greater diameter at the top and bottom of the combination, the whole being that two horizontal cylinders are joined at many points by rings in vertical planes, the cylinders being at the top and bottom of the vertical diameter of the rings. The fire was made in the inside of the arrangement resting on the rings as grate bars, and the feed water introduced into the

# MISCELLANEOUS BOILERS, ETC.

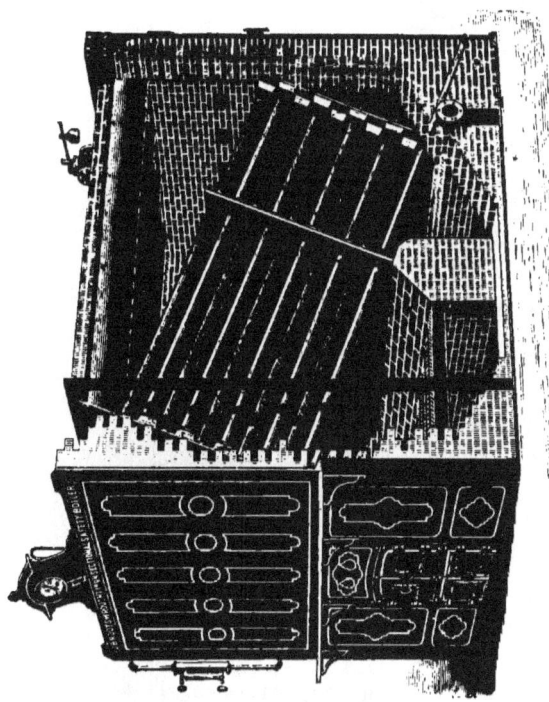

ROOT'S NEW SECTIONAL SAFETY BOILER.

## BOILER OF THE STEAMER "ANTHRACITE."

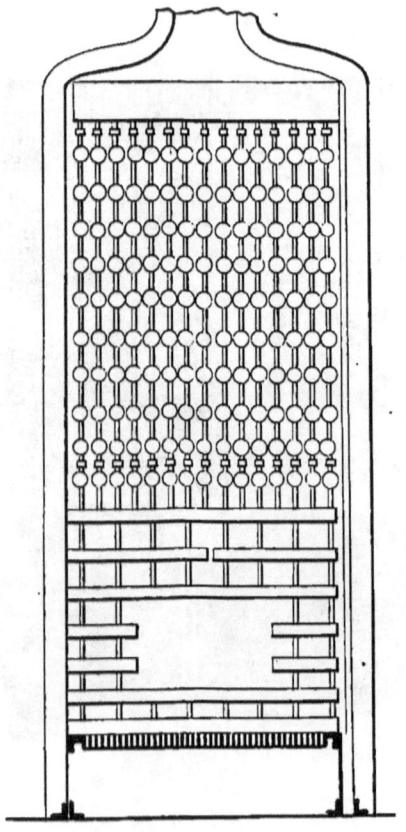

**THE FIRMENICH BOILER.**

lower straight pipe branching to the ring pipes, and being evaporated therein, and steam taken from the upper straight pipe.

Of course with poor water a steam drum above had to be provided so that the water line came above the rings and a circulation either through the pipes or external pipes added therefore.

With the larger Belleville boilers the feed is introduced into a vertical drum and then crosses the steam drum in a kind of trough to the vertical mud-drum, whence it passes down to the distributing drum. With the exception of the passage across the steam drum whereby the water gets hot enough to render the sulphates and carbonates of lime insoluble, and the addition of a circulating pump, the arrangement is identical with that of Benson. Belleville circulates by gravity and the claim is that from four to eight times the water evaporated is circulated. Heine also uses the enclosed feed heater in the same way, only the outside mud-drum is not used, the blow-offs being taken from the enclosed chamber.

We give a sketch of Ogle's boiler with cuts of Heine's as built at St. Louis; the latest type of Belleville's and Root's, and a sketch of one of Herreshoff's smaller boilers,—the feed being into the outside of the flat top coil, and then down the spirals towards the fire and then out to the separator. In many cases a turn or two of pipe in the furnaces is added after the steam leaves the separator, to dry or superheat the steam. A second coil of spiral pipe was added on the outside of the "beehive coil" in order to increase heating surface, and other coils have been added till one of the latter boilers is as follows: A coil of small pipe fed at the bottom passes up close to the sheet-iron shell, three flat shells come next, followed by a single "beehive coil." The Latta boiler, we have already illustrated, and we add a cut of Firmenich's. Belleville's and Root's boilers have been very extensively used in certain localities. Herreshoff and Latta have had great success with special applications—one to steam yachts and launches, the other to fire engines.

Closely allied is the boiler now used by Loftus Perkins, the descendant of Jacob Perkins, who left Newburyport, Mass., for England, and who was the originator of many things since claimed as novelties, some unsuccessful, as the steam gun which he used with the spiral pipe boiler, some since successful, as the soft iron toothless disc now used for cutting steel rails, cold, which he, however, only tried on files hardened.

The boiler used on the steamer "Anthracite," consisted of 140 horizontal pipes in 10 layers, 14 in a layer horizontally, and 10 in a layer vertically. These are united vertically by short nipples of small pipe, and at the top each of the 14 pipes of the top layer is connected at the middle of its length to a transverse pipe of larger diameter, acting as a steam drum. The furnace is made of seven rows of pipes across the back transversely to upper group, seven rows on each side, and completed rows across the front two rows being separated in the centre, and two shortened rows leaving an opening for fire door, The boiler is only used with distilled water, and the first thought we have is, that with the smallness of the rising pipes and the lack of special provision for circulation or return of water

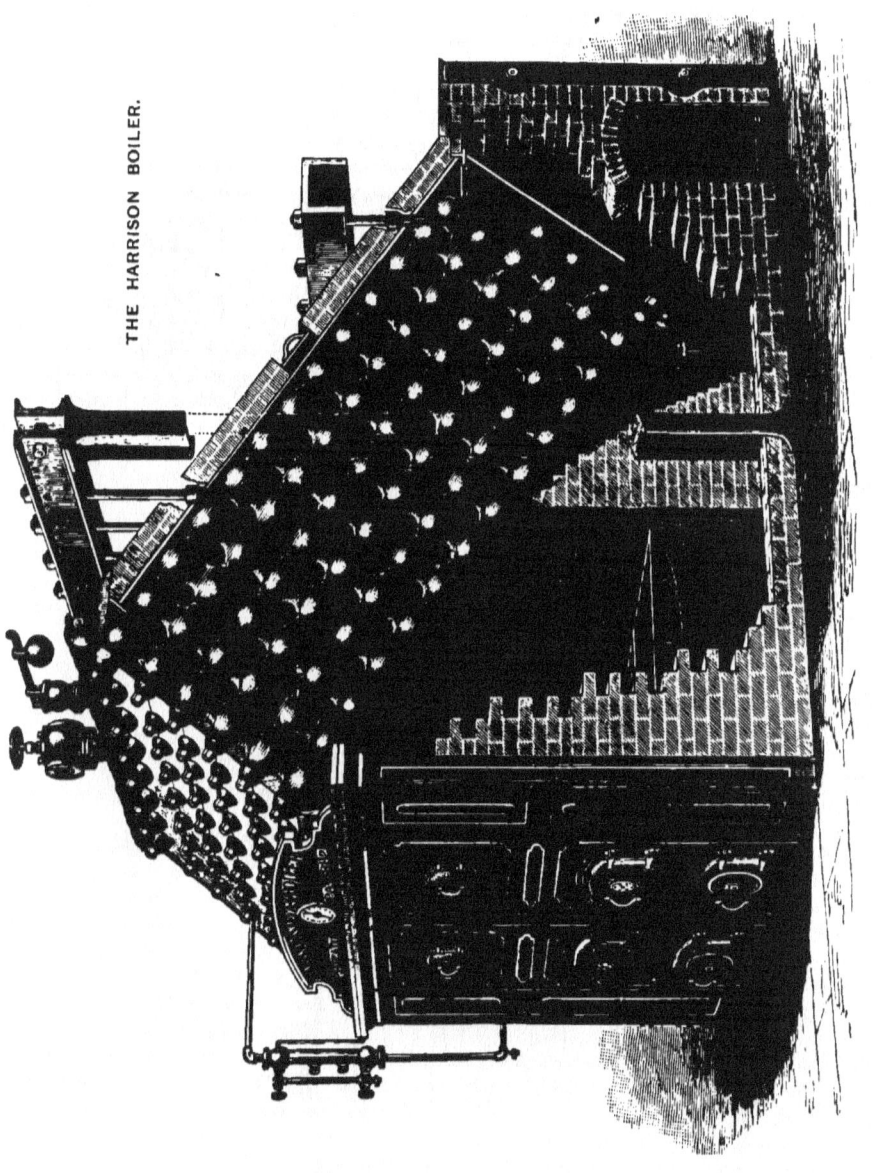

THE HARRISON BOILER.

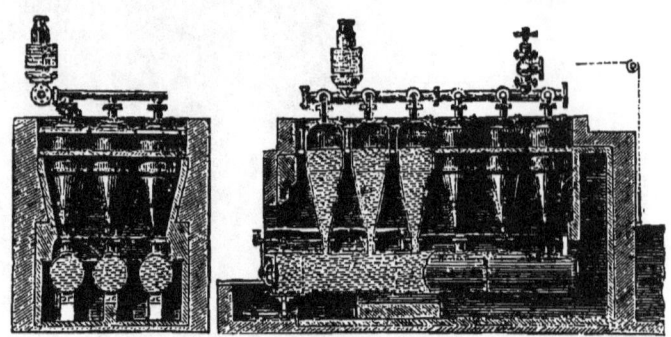

SHEPHERD'S BOILER.

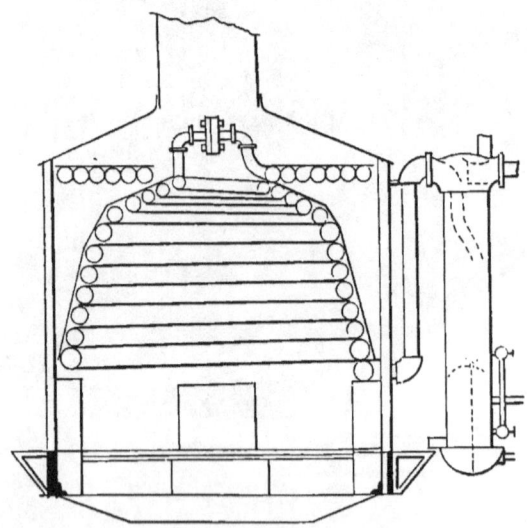

THE HERRESHOFF BOILER.

to the tubes around the furnace, we should expect a very marked "foaming" which was found to be the case at the trial made by a Board of United States Engineers, United States Navy.

We also give a sketch of an element of section of Kelly's boiler.

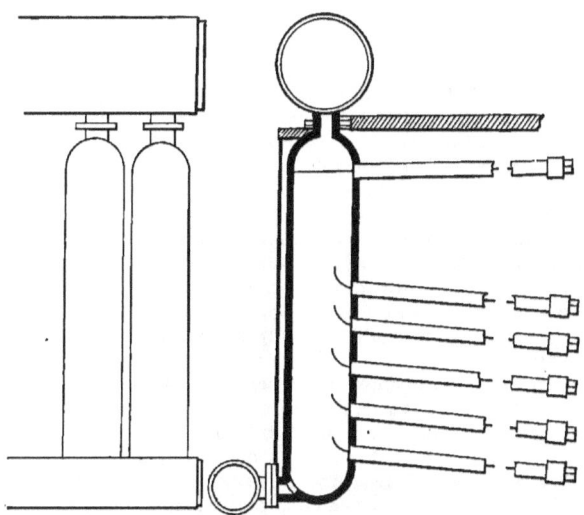

ELEMENT OF A SECTION OF KELLY'S BOILER.

Hardly to be distinguished from the water tube, or tubulous boilers, come the sectional boilers, such as Shepherd's and Cadiat's, already mentioned, the most prominent among them being the Harrison, which is made of cast iron globes united in straight lines by necks, flanges and bolts. Each line of globes is thus a straight tube with alternate enlargements and contractions, and the boiler is subject to the incidentals of all the others of the tubulous class. The joints are, of course, much more numerous than with wrought-iron tubes, but they are claimed to be easier to make and to remain in better order. A great number of these boilers have been in satisfactory use in the eastern part of the United States, and elsewhere.

We may then sum up as follows: After a desirable, and counting all things, an economical evaporation has been decided upon, and the grate area defined from the fuel consumption, the heating surface may then be computed and the results compared with the available experience. That form of boiler which can be constructed and set in operation for the least money, without lowering the quality of the material and workmanship, and without interfering with the quality of the fuel and water,—in brief its durability,—is the one to be recommended.

### APPURTENANCES TO A BOILER.

In operating boilers a multitude of fittings and tools are required. First, as to the furnace. The fireman has to use a shovel for coal and for shavings a kind of push hoe; for wood his hands only in getting the fuel on to the fire. The choice of a shovel depends much on the man, but a broad flat one seems favorable to spread the fuel uniformly; as the coal is thrown forward the shovel is struck on the threshold of the fire door in a peculiar manner which jerks the coal upward and assists in the operation. The other fire tools are the poker, the rake, a kind of hoe, a slice bar for running up through the grate, and a flat bar for lifting from above the grate; the others are added as circumstances require. A shovel for removing ashes will probably be required in addition to that used for firing.

To regulate the air supply a damper must be used, placed on the door in the flue or in the smoke connection, and this is either worked by hand, or automatically. The fire door is usually supplied by a register, which is sometimes arranged to be set wide open by the act of closing the door, when a weight falling against a dash pot gradually closes it again. This was one of the earliest attempts to prevent smoke. A swinging plate damper in the stack is sometimes connected with a loaded diaphragm, or to a piston in a cylinder working against a spring, or even to a common safety valve lever. When the pressure of steam rises, the movement of the diaphragm or piston, suitably multiplied, turns the damper and reduces the draft. As the pressure falls the damper is opened by the reverse action.

In supplying feed water to the boiler a check-valve is placed conveniently close to the shell, or point of attachment, for the point of attachment of the feed pipe is peculiarly liable to be strained by changes of temperature of the feed water as well as the fire, consequently an extra liability to leakage exists; and from the fact that this connection in a stationary boiler is usually in the furnace, a great deal of corrosion takes place, and is to be expected, at this part. The joint is usually made by screwing the pipe into a flange secured to the shell by bolts or rivets. If anything happens to the feed water apparatus, such as the leakage of a valve, or anything which requires disconnecting the feed pipes, this check valve enables it to be done without delay. It also acts as a guard and relief to the discharge valves of the pump.

# MISCELLANEOUS BOILERS, ETC. 185

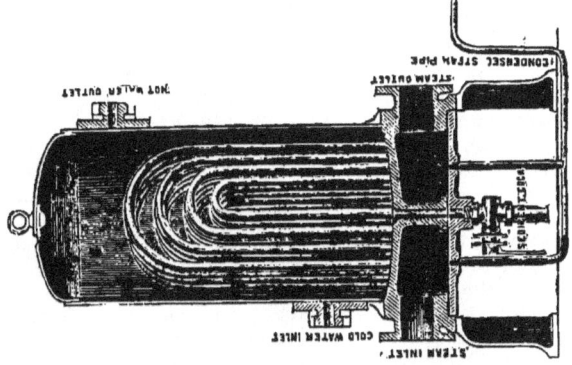

FEED WATER HEATERS.

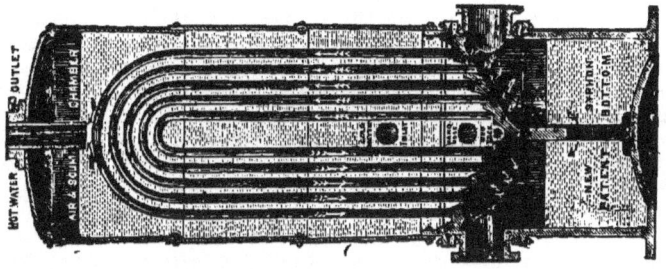

When a feed-water heater is used, it may be placed between the pump and boiler, or the pump may be placed between the heater and boiler; in the former case any ordinary form of feed pump may be employed, but the feed water must be separate from the exhaust steam used for heating, for the feed water must be at the boiler pressure, while the exhaust steam is of course near the atmospheric pressure. By using a jet of fine steam to raise the upper portion only of the water which is in the heater, the lime salts may be rendered insoluble, and can be caught in a quiet pan, or blown off at the surface. This is Strong's heater. No economy of fuel is obtained by the use of live steam, its effect is the purification of the water. On the other hand, if the feed water pass first into the heater, any kind of a heater may be employed; but the pump valves must be prepared to take hot water, and it is desirable to have the feed water supplied to the force pumps from a higher level, so that no attempt at suction may be required, for the water gives off steam in large volumes at a very slight reduction below the pressure of the atmosphere, and, filling the pump chambers, interferes with its satisfactory working.

We give a few forms of heaters used, but do not think it advisable to enter more fully into the subject. In a great portion of the western practice a closed cylinder set horizontally is used, which is filled with the feed water to about half; the exhaust steam from the engines is then passed over the surface of the water, which is held by blades from being taken along with the steam, for the exhaust steam often leaves the cylinder at a pressure of 75 pounds to the inch. With the heater just described two sets of pumps are employed, one to lift from the ground, or well, to the heater, a work which may be omitted if the supply be high enough, and the other a force pump, filling by gravity from the heater, usually placed above the pumps, and forcing into the boiler. The favorite western arrangement is a beam, crank and fly-wheel engine working two lifting pumps on one side, and two force pumps on the other side of the beam, the feed being carried through the hollow columns which support the heater. The whole arrangement goes by the name of "doctor." Although the machine can hardly be surpassed for general ugliness of appearance, yet it is eminently adapted to the work to be done, and all attempts to drive it from the boats on the river, and its vicinity, have failed, for it represents a long experience with a peculiar set of conditions.

The feed pumps used are either attached to the main engine, or driven by an independent steam engine, and in no one point of practice is there more universal disagreement on the two sides of the Atlantic. In England and Europe the attachment of the pump to the engine is a matter of course, and even the smallest engines are provided with them: it is only in large engines, driving mills, or where a number of engines are employed, or the water is used for something else, that an independent pumping engine is used. The argument, which can not be contradicted, is, that it costs less steam, hence fuel, to do the work of pumping, is but a small addition to the work of the engine, $\frac{35}{100}$ per cent. if the water used be 30 pounds per horse power per hour, and the steam pressure 100 pounds above the atmosphere, and varying directly with the steam pressure and the economy of the engine—than by a small, non-expanding, slow working engine. In the United States this was also the practice, but at present the use of the independent engine is universal, and it is beginning to be introduced into England from this country. The chief merit is in its being able to run when the engine is shut down, and to vary the quantity of feed while the uniformity of speed of the engine is not interfered with. The former adds to the security of the boiler if the engine be suddenly stopped, for the water supply will not stop with a big fire, when of course the supply should be moderated, but from being almost the first thing to be looked after it becomes a very secondary matter. As in a good engine a horse power can be furnished from the boiler at a cost of from 20 to 30 pounds of water per hour, while with the small one, as frequently met with, this cost amounts to from 60 to 100, or say, three times as much for the independent engine. Let us see what this gives: With a boiler evaporation of 6 pounds of water for 1 pound of steam, we see that it might take 17 pounds of fuel for the worst case we have supposed against 3 for the best engine, a difference of 14 pounds per hour, or 140 pounds per day for each horse power required. If the steam pressure is 100 pounds we have 20 lbs. × 230 feet, = 4,600 foot- pounds per horse power per hour, and if the engine indi- cates 100 horse power, we shall have 460 × 100, or 460,000 foot- pounds, as the energy required to pump the feed for an hour. Now, a horse power is 550 foot pounds ex- erted for a second, or 33,000 a minute, or 1,980,000 for an hour. This is about one- fourth a horse power, which would require about one-fourth of the 140 pounds, 35 pounds, of coal a day, assumed to be the greatest difference in cost likely to be met with, or say 40 as a limit, an expense

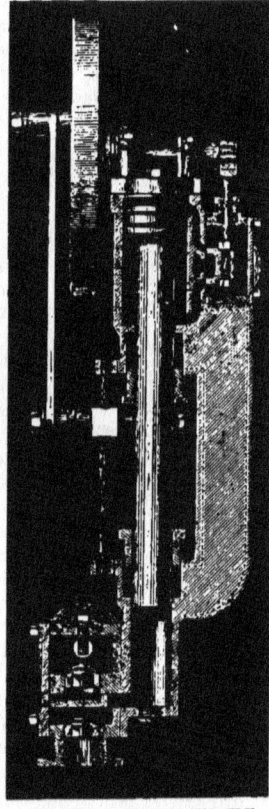

SECTION OF CRANK, AND FLY WHEEL FEED PUMP.

ranging from 3 to 25 cents a day, according to the price of fuel.

If, on the other hand, the main engine requires 40 pounds of water per horse power per hour, the quantity required for our 100 horse engine, with 100 pounds of steam, will be twice as great, or about one-half a horse power; and if now we have a feed pump of proper proportions, we have, for half a horse-power, 30 lbs. of water, and for a boiler evaporation of 9 in place of 6, 3.3 pounds of coal, used by the independent engine. An engine using 20 lbs. of water gives the same result, viz., 11 lbs. per day difference in place of 40 lbs. The convenience and security of a separate pump would seem to far outweigh this slight economy.

Of the multitude of steam pumps made we find two classes, the direct acting and the crank and fly wheel. The latter are more economical in using steam, as the steam may be used expansively, but it may be doubted if the additional couplings caused by the greater number of connections do not overbalance this, because, of course, the economy claimed can not reach that which we have just investigated, where the pump was attached to the main engine.

With the choice of a boiler feed pump we shall have little to do. The cost, and cost of repairs, or collectively, the annual cost of the pump is the item to be gov erned by, and this varies necessarily in different localities and at different times. This is not always an easy matter to do, but we have seen that the difference among them must be of comparative unimportance, and we shall dismiss the subject without attempting to make a choice among the many excellent machines in the market.

We illustrate only one form of pump, not made in this country, as simple and seeming well adapted to any work.

The use of a jet or line of motion of another stream of fluid across the stream of fluid to communicate motion from the first stream to the second stream has long been known. A stream

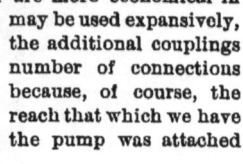

ELEVATION OF CRANK AND FLY-WHEEL FEED PUMP.

of water acting at high pressure, 400 lbs., in a 2-inch pipe upon a second stream, had the power to lift a column of water 120 feet in a 6-inch pipe, bringing with it an immense quantity of sand and 'gravel. We have here a convincing example of one stream contributing enough of its energy to move the other two, one of water and one of sand, while the resulting velocity is that due to the energy of the first stream acting upon the whole mass. A second example, which we will discuss later, is a jet of steam to set a current of air in motion.

The use of a jet of steam for moving a stream of water originated the steam syphon, and various other forms of the steam jet pump, a useful adjunct to locomotives in a new country.

The energy of motion of a body is well known to be the product of its mass by the half square of its velocity, hence it is possible to communicate to a body of little weight a large amount of energy by moving it fast enough, and in fact the energy of motion would only be limited by

## MISCELLANEOUS BOILERS, ETC.

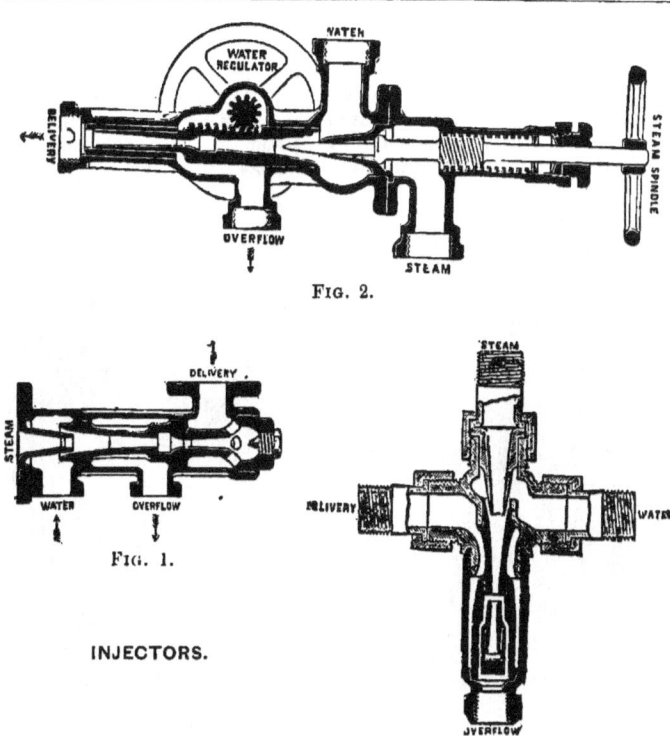

FIG. 2.

FIG. 1.

INJECTORS.

FIG. 3.

the speed which can be given the body. In this way a small weight of steam flowing from an orifice into a properly shaped jet of water is condensed while the velocity of the steam is greater than if flowing into air; the energy thus communicated is made sufficiently great by increasing the weight of steam, which can be done by increasing the area of the steam way, until we find such jet pumps adapted to many purposes. There are, however, two which are of interest to us in this connection, or rather only one with a modification. The well known injector invented by M. Henri Giffard, and its large family of lifting and non-lifting varieties, all differing in details as to form of nozzles, area of passages, distances between nozzles, and that class of instruments in which, after a certain energy and velocity have been reached, the operation is repeated. These might be called "consecutive" instruments. Our illustration, Fig. 1, shows one of the simplest non-adjustable kinds. Within a few years this principle of increase of energy by increase of mass or velocity has been applied by increasing the mass of steam used until we find that not only can a few pounds weight of steam put into a boiler a good many more pounds of water at a much higher temperature than it had, but that in a non-condensing engine it is possible, by using the exhaust steam in part, to put into the boiler at a much higher pressure and temperature, a weight of water which is still greater than that of the steam moving it.

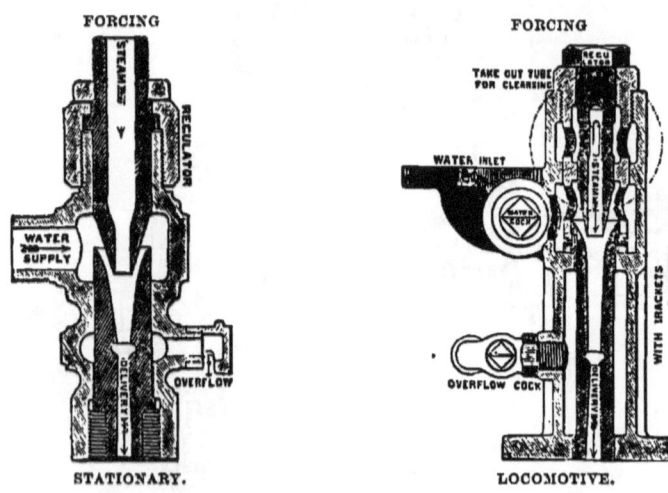

INJECTORS.

When the injector first made its appearance, it was, by many, considered as almost a paradox, especially by those who looked at the question as one of hydrostatics only. That steam from a boiler could put water back into it at the same pressure, and overcome the friction of the passages without the aid that a steam pump had of a difference of piston areas, was to them a puzzle. The use of exhaust steam at atmospheric pressure for the purpose of putting water into a boiler at a pressure of 150 pounds per square inch would be to such minds utterly incomprehensible. We think there are two classes of instruments made for the difference in steam pressures below and above 60 pounds, differing in area and disposition of the passages.

As this modification is one of the latest, so it is not one of the most important, for we have seen that the cost of the feed water put into the boilers was from one-fourth to one-third of 1 per cent. Now, a horse power with the best engine and boiler, non-condensing engine, runs from 2½ to 3½ lbs. of coal per hour, a value by the way corresponding to an average between the ¼ and ⅓ of 1 per cent. given above, the former being taken for a condensing engine using 20 lbs. of water, the latter, 25 lbs., giving with 10 for evaporation, 2½ lbs. of coal per hour per indicated horse power. If, now, the engine runs 500 indicated horse power 24 hours in the day, we have $\dfrac{24 \times 500 \times 7 \times 5}{2400 \times 2}$. The quantities are then,

24 hours per day.
500 horse power.
$\dfrac{7}{2400}$ the average between ¼ and ⅓ per cent.
$\dfrac{5}{2}$ the 2½ lbs. of coal required for one horse power.

This costs from 5 to 25 cents per day, or from $15 to $75 per year, the feed water being delivered at a temperature of about 150° F., and the cost of a heater being saved in addition. We see, therefore, that the feed water can easily be put into the boiler, and, as we have seen elsewhere, all that can be saved from the expense is so much net gain, while it is evident from our investigation, that more can be saved by the use of a first-class heater, leaving the water, say at 200° F., and then taking out what is used by the pump, than by an exhaust injector, as this modification is known.

The use of an injector has this to recommend it, that the feed water can not be introduced into the boiler cold or nearly so, but must be warmed by contact with the steam, and the value of this has been already shown. In small boilers where no heater is used an exhaust injector is better than a pump, and so is an ordinary injector; but the former includes in itself an exhaust heater, saving a portion of heat from the exhaust, besides taking the power as heat also, while with the common injector the heat for power and raising temperature are both derived from the live steam in the boiler. The latter portion of heat is, of course, directly returned to the boiler without loss, but that for power is necessarily expended. As to the amount of power used by pump and injector compared with each other, it would seem that the pump is most efficient. There have been many comparative trials of pump and injector, but the results have usually been unsatisfactory from contained discrepancies. We may, however, sum up our impressions as follows:

For engines with condensers, either a pump or common injector may be used.

For engines with jet condensers, an injector is to be preferred to a pump, as the temperature of the feed is necessarily much higher than the hot well. This preference is not because of economy of fuel but durability of the boiler.

For non-condensing engines, in order of choice, a pump and first-class heater, an exhaust injector, a common injector.

The exhaust injectors have not yet been used enough to develop their full capabilities, but there is no reason to doubt that in a few years the same confidence should be felt in them as in the older forms.

We illustrate only a simple form of injectors, in sections; these are not made in this country.

The application of a steam jet to induce a current of air for draft is nearly as old as the locomotive with which it originated and to which its use now is almost restricted, and to boilers of the same class where a sudden call for steam can be rapidly met. In the most simple form a pipe is led from the boiler to the stack, if of iron, if not, to some of the flues or tubes, which is terminated by a reducer with short nipple; a 1″ pipe with a ⅜″ or ¼″ nipple, sufficing to raise the gauge from 20 to 90 ℔s. in 7 minutes for a 100 horse power engine.

The chimney used in this case was 18 inches diameter and 25 feet high.

For plain jets and nozzles of this class in open cylinders, Mr. J. A. Langridge, Member Institution of Civil Engineers, concludes essentially:

1. The action is due to the friction of one fluid on the other, and that by dividing the jets the surface of contact of the fluids is much increased for the same masses of fluid; or the same draft may be produced with less steam.

2. The effect of the draft is increased by lengthening the chimney, but the effect is smaller from four diameters to eight diameters than less than four diameters. Above eight there is a falling off.

3. He states that the draft measured in inches of water, inches of diameter and pounds per square inch above the atmosphere, may be computed as follows:

Draft equals 37 times the fifth power of the third root of the diameter of the blast pipe, times the fourth power of the fifth root of the pressure, divided by the square of the diameter of the chimney.

Very much better results are obtained by giving proper form to the nozzle and guiding surfaces around the jets.

Blow-off valves are used at the bottom and surface; the former are used intermittently, and as their use includes running out all the water in the boiler when it is desired to remove it, they are quite large, and should, as before noted, have sliding gate with room for mud. The attachment to the shell of such a large pipe should be by flange. The upper blow-off usually takes its water from such a place in the boiler that a well defined down current follows a horizontal one. At this angle the water is comparatively quiet. Another method is to provide an inside pan where the water is shielded from upward streams and bubble, where the scum on the surface may have a chance to form. Sometimes a drum outside the boiler is connected at two points with this pan and the circulation set up by differences of temperature brings the water out with its impurities; the latter have time to settle, or if not cooled sufficiently, remain at the surface and are there removed. The action of a surface blow-off may be intermittent or constant. In the latter case a loss of heat occurs which we have fully discussed in this chapter.

Gauge cocks are put in at different levels near the water line. The lowest is usually put in so that a full gauge of water lies over the danger point, or highest metal exposed to the direct action of the hot gas on the

COMBINATION GAUGE.

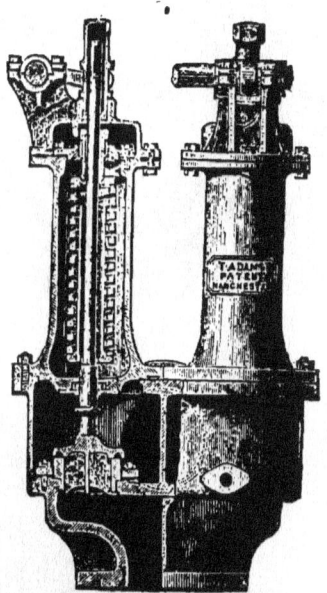

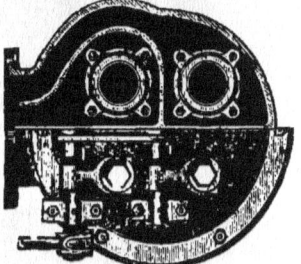

TYPE OF SAFETY VALVE FOR MARINE BOILERS.

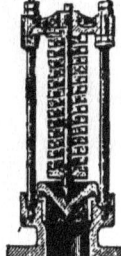

ORDINARY SAFETY VALVES.

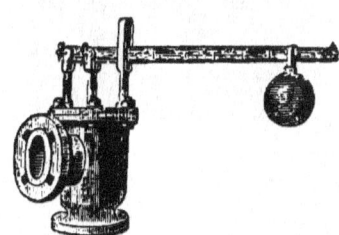

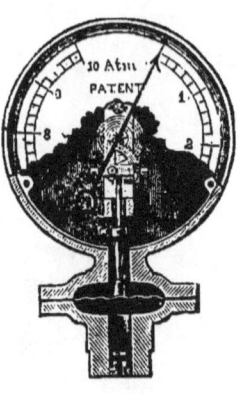

STEAM GAUGES.

FORGED WROUGHT IRON MAN HOLE AND COVER.

CORRUGATED FLUE.

second return thereof. The cocks are in number from two up, three, or four in all, being the common number. The upper one is placed at as high a level as it is thought can be used without foaming. In addition, one or two glass tubes are sometimes used. The brass fittings in which the latter are inserted should be provided with four valves, one between the glass and boiler at each end of the tube, and one at each end in the line of the tube, so it can be cleaned by washing from either end and a rod can be run through it. The tube is packed in place by gum washers and double nuts. Specially soft glass has to be used, and great care taken not to scratch the glass, or a break is sure to happen. By shutting off the glass from the boiler it can easily be replaced.

We illustrate a combined gauge glass and gauge cocks.

A float inside the boiler attached to two arms on a spindle passing through the head of the boiler in a properly packed box, varies with the water and shows its position by a needle attached to the spindle. The constant fluctuations of these instruments show them to be in working order, but the gauge cock should be used every hour.

Safety valves are loaded by dead weight, weight and lever, direct springs, and spring and lever. The valves are usually plain cones, but of late years, the portion beyond the cone has been modified in a manner easily seen from the illustrations.

The most important adjunct of a boiler is the pressure gauge.

Pressure gauges are made either with diaphragm, as a spring against which the steam presses, or by a flattened curved tube which tends to become circular in section with increase of pressure.

The last adjunct of a boiler is the manhole and its cover and the hand holes. The common form is too well known for description, but we give an illustration of a forged wrought iron fitting, with cover bolted on a ground joint. In this country such a forging would be found difficult, but we have a few made of gun metal, or "gun" cast iron.

In conclusion we give, in addition to the many cases among our boiler illustrations, a drawing of one of the corrugated flues, or furnace tubes.

www.ingramcontent.com/pod-product-compliance
Lightning Source LLC
Chambersburg PA
CBHW020858230426
43666CB00008B/1224